Malaria

Host Responses to Infection

Editor

Mary M. Stevenson, Ph.D.
Associate Professor
Department of Medicine
and
Centre for the Study of Host Resistance
McGill University
The Montreal General Hospital Research Institute
Montreal, Quebec, Canada

CRC Press
Taylor & Francis Group
Boca Raton London New York

CRC Press is an imprint of the
Taylor & Francis Group, an **informa** business

First published 1989 by CRC Press
Taylor & Francis Group
6000 Broken Sound Parkway NW, Suite 300
Boca Raton, FL 33487-2742

Reissued 2018 by CRC Press

Publisher's Note
The publisher has gone to great lengths to ensure the quality of this reprint but points out that some imperfections in the original copies may be apparent.

Disclaimer
The publisher has made every effort to trace copyright holders and welcomes correspondence from those they have been unable to contact.

ISBN 13: 978-1-138-10606-2 (hbk)
ISBN 13: 978-1-138-56060-4 (pbk)
ISBN 13: 978-0-203-71323-5 (ebk)

Visit the Taylor & Francis Web site at http://www.taylorandfrancis.com and the
CRC Press Web site at http://www.crcpress.com

to
F.R.S.

PREFACE

Malaria is the number one infectious disease threatening mankind today. This may come as a surprise to most inhabitants of the western world, particularly to North Americans, since we are bombarded daily, via the media, concerning the threat of AIDS, which is indeed a major and serious health problem. However, about half of the world population lives in areas, primarily in underdeveloped nations, where the incidence of malaria is either at or reaching epidemic proportions. The overall world malaria situation has not improved since 1970. There are an estimated 500 million acute cases per year, resulting in the death of 3 to 4 children per minute in Africa alone. The threat of malaria continues, despite control measures, due to the emergence of drug-resistant strains of the *Plasmodium* parasite, the causative organism of malaria, and the emergence of insecticide-resistant vectors, mosquitoes of the genus *Anopheles*.

Genetically engineered vaccines have recently been developed for use against two distinct developmental stages of *P. falciparum*, the major causative species of human malaria. These vaccines, which are against either the sporozoite, the stage of the parasite injected by the mosquito, or the asexual erythrocyte stage or merozoite, which causes the symptomatology clinically associated with malaria and can be fatal, are currently undergoing trials in human volunteers. However, for malaria, as for other major infectious diseases, including leishmaniasis, schistosomiasis, trypanosomiasis, and AIDS, the exact mechanisms which the host uses to limit invasion of the microorganism to control its multiplication and to eliminate it have not yet been defined. In each case, studies of immune responses to infection in man and experimental animals have provided information concerning these mechanisms. For example, there is evidence in the literature that among humans in endemic areas recovery from infection leads to immunity. This volume attempts to provide further insight into the underlying mechanisms of this immunity. Because it is the multiplication of the *Plasmodium* parasites within the host erythrocyte which causes the clinical symptoms of malaria, this volume focuses on recent experimental evidence concerning the mechanisms of resistance used by the host in response to the asexual erythrocyte or blood stage of malaria. Although the data are primarily derived from experimental murine models, data derived from human studies are also included.

In the 1930s, experimental evidence was published that supported a role for both humoral immunity (Coggeshall, L. T. and Kumm, H. W., Demonstration of passive immunity in experimental monkey malaria, *J. Exp. Med.*, 66, 177, 1937) and cell-mediated immunity (Cannon, P. R. and Taliaferro, W. H., Acquired immunity in avian malaria. III. Cellular reactions in infection and superinfection, *J. Prev. Med.*, 5, 37, 1931) in resistance to the blood stage of malaria. In each case, these mechanisms were postulated to lead to control of the infection and elimination of the parasite and recovery of the host. Since that time much experimental effort has been directed towards elucidating the role of antibodies in protective immunity against malaria. From the 1930s to the 1970s, antibodies were assumed to be the primary mediator of resistance to the erythrocytic stage of malaria; the role of cell-mediated immunity was thought to be minimal. Since the 1970s however, our knowledge of cell-mediated immunity, particularly the role of, and interactions between, T cells and macrophages and their respective cytokines in host resistance against infection has expanded rapidly. The pendulum has, therefore, swung back, and the role of cell-mediated immunity in protective immunity against this stage of malaria is once again receiving much deserved attention. Both mechanisms of immunity are now under active investigation. This suggests that, at last, a balance has been achieved and that we have come full circle in accepting the concept that both humoral and cell-mediated immune mechanisms play a role in protective immunity against malaria.

In the case of blood-stage infection with some murine malaria species (*P. chabaudi adami, P. chabaudi chabaudi, P. vinckei petteri*) but not others (*P. yoelii*) B cell deficient mice have been shown to resolve acute infection and to resist reinfection (Grun, J. L. and Weidanz, W. P., Immunity to *Plasmodium chabaudi adami* in the B-cell-deficient mouse, *Nature*, 290, 143, 1981, Malancon-Kaplan, J. and Weidanz, W. P., Chapter 3). This observation suggested for the first time that a nonantibody, T cell dependent, cell-mediated immune mechanism can function in resistance to blood-stage malaria. Although B cell deficient mice can resolve an acute infection with these parasites, the animals develop a chronic low-grade parasitemia, thus indicating a role for antibodies in complete elimination of the infection. Taken together, these observations in B cell deficient mice support the hypothesis that both humoral and cell-mediated immunity function in the resolution of blood-stage malaria. It is of interest to point out that in order to achieve effective vaccination against sporozoites of *P. falciparium* the immunogen must contain not only B cell epitopes but also T cell epitopes for the induction of cell-mediated immunity. This is an indication of the importance of both humoral and cell-mediated immunity in the response of the host to all stages of malaria.

This volume consists of seven chapters which address the question of how a host responds to a blood-stage malaria infection. Since malaria is a complex disease, the host can respond in varied and complex ways which involve both innate resistance and acquired or immunologically specific immune responses.

The book begins with a chapter on humoral immunity to malaria. This is followed by two contributions describing the role of cell-mediated immunity in response to malaria. The first deals with cell-mediated immune responses during a primary, acute infection. It discusses the role of T cells and the activation of effector cells including NK cells, polymorphonuclear leukocytes and macrophages, as well as the role of the spleen in cell-mediated immunity to malaria. The second of these chapters discusses the role of cellular immunity mediated by T cells and effector cells, including macrophages, neutrophils, eosinophils, endothelial cells, and NK cells, with special emphasis on enhancement of this mechanism by vaccination. The next two chapters examine in detail the effector cells and molecules of the nonantibody-dependent mechanisms leading to the intraerythrocytic destruction of *Plasmodium* parasites: the role of macrophages in resistance to malaria, and crisis form factor, which has been described in humans. The sixth chapter discusses the mediators of inflammation, which are primarily derived from macrophages, as the etiological basis of the immunopathology of malaria. The seventh and final chapter discusses the topic of genetic control of host resistance to malaria. Each chapter includes a section on the extremely important topic of strategies for antimalarial vaccination from the perspective of the subject matter of that chapter.

I would like to thank the contributors for the care with which they prepared the manuscripts and promptness with which they were submitted. I am indebted to my mentor and colleague, Dr. Emil Skamene, for introducing me to the concept of genetic control of host resistance to infection.

Mary M. Stevenson

THE EDITOR

Mary M. Stevenson, Ph.D., is an Associate Professor in the Department of Medicine and an Associate Member of the Department of Physiology, McGill University, Montreal. She is a member of the McGill Centre for the Study of Host Resistance and the University Medical Clinic, which are part of the Montreal General Hospital Research Institute.

Dr. Stevenson received her B.A. degree from Hood College, Frederick, MD, in 1973. She obtained her M.S. and Ph.D. degrees in 1977 and 1979, respectively, from the Department of Biology, Catholic University of America, Washington, D.C. She conducted her M.S. and Ph.D. thesis research at the Biology Branch, National Cancer Institute, National Institutes of Health, Bethesda, MD. After completing her postdoctoral studies in 1981 in the Division of Clinical Immunology and Allergy at the Montreal General Hospital, Dr. Stevenson was the recipient of a Foundation Award from the Montreal General Hospital for the year 1981 to 1982. During this year she was a Research Associate at the Montreal General Hospital Research Institute. In 1982, she was appointed to the Department of Medicine of McGill University as an Assistant Professor.

Dr. Stevenson was awarded a Scholarship from the Medical Research Council of Canada in 1982 and has been the recipient of research grants from the Medical Research Council of Canada, Fonds de le Recherche en Sante du Quebec, the National Cancer Institute of Canada, and the Thrasher Research Fund.

She is a member of the Reticuloendothelial Society, the American Society for Microbiology, the American Association of Immunologists, the Cancer Research Society of Montreal and the Canadian Society for Immunology. She is the author of more than 40 publications. Her current major research interests include genetic control of host resistance to infection with intracellular parasites, particularly the murine malaria species *Plasmodium chabaudi* AS, and the role of cell-mediated immunity in host resistance to malaria.

CONTRIBUTORS

Geeta Chaudhri, Ph.D.
Department of Experimental Pathology
John Curtin School of Medical Research
Canberra, Australia

Ian A. Clark, Ph.D., D.Sc.
Principal Research Fellow
Department of Zoology
Australian National University
Canberra, Australia

James B. Jensen, Ph.D.
Professor
Department of Microbiology and Public
 Health
Michigan State University
East Lansing, Michigan

K. Rebecca Jones, B.D.S.
Research Assistant
Department of Immunology
University College
and
Middlesex School of Medicine
London, England

Johanne Melancon-Kaplan, Ph.D.
Research Assistant Professor
Department of Microbiology and
 Immunology
Hahnemann University School of
 Medicine
Philadelphia, Pennsylvania

**J. H. L. Playfair, MB.B.Chir., D.Sc.,
 Ph.D.**
Professor
Department of Immunology
University College
and
Middlesex School of Medicine
London, England

Hannah Lustig Shear, Ph.D.
Research Associate Professor
Department of Medical and Molecular
 Parasitology
New York University Medical Center
New York, New York

Mary M. Stevenson, Ph.D.
Associate Professor
Department of Medicine
and
Centre for the Study of Host Resistance
McGill University
The Montreal General Hospital Research
 Institute
Montreal, Quebec, Canada

Diane Wallace Taylor, Ph.D.
Associate Professor
Department of Biology
Reiss Science Center
Georgetown University
Washington, D.C.

Janice Taverne, Ph.D.
Research Associate
Department of Immunology
University College
and
Middlesex School of Medicine
London, England

William P. Weidanz, Ph.D.
Professor
Department of Microbiology and
 Immunology
Hahnemann University School of
 Medicine
Philadelphia, Pennsylvania

TABLE OF CONTENTS

Chapter 1

HUMORAL IMMUNE RESPONSES IN MICE AND MAN TO MALARIAL PARASITES

Diane Wallace Taylor

TABLE OF CONTENTS

I. INTRODUCTION

The great debate continues! Today, it is clear that antibodies (Abs) play a role in protective immunity against malarial parasites, but really how important are they? Originally, immunity to malaria was thought to be mediated solely by phagocytic cells, but at the end of the 1930s, serum containing antimalarial Abs was found to transfer protection to naive animals.[1,2] Since then, Abs have been considered to be the primary immunologic defense of the host against the parasite. Recent studies on monokines and T cell-derived lymphokines, however, have led some investigators to suggest that we may need to reevaluate our thinking. It, therefore, seems timely to review information on Abs and their significance in malarial immunity.

Malaria is a complex disease, with the parasite going through different developmental stages, e.g., sporozoites, exoerythrocytic forms within hepatocytes, and blood stage forms (including intraerythrocytic rings, trophozoites, schizonts, gametocytes, and extracellular merozoites). It is now evident that the effectiveness of Abs differs in eliminating the various parasite stages. An additional level of complexity emerges when one realizes information on humoral immunity was acquired using different host-parasite systems; that is, by studying strains of mice infected with one of four different species of murine malaria (*Plasmodium berghei*, *P. yoelii*, *P. chabaudi*, and *P. vinckei*); rats with *P. berghei;* different species of monkeys parasitized with at least four primate malarial species (*P. knowlesi*, *P. inui*, *P. fragile*, and *P. cynomolgi*); and man infected with four species of human parasites (*Plasmodium falciparum*, *P. vivax*, *P. ovale*, and *P. malariae*). The pathology of the disease differs among the parasite species, and the same parasite may produce a different course of infection in genetically different hosts. Thus, both the kinetics and nature of Ab production and the protective value of Abs may vary with the host, the species of parasite being studied, as well as the stage of the parasite.

Although a functional role for Abs in malarial immunity is well established, it is unclear how Abs mediate their protective affects. (The exception to this may be Abs to gametes discussed at the end of the chapter.) Concepts on the effector functions of Abs have been proposed, but few conclusively demonstrated.

Surprisingly, little is known about the immunologic induction of Ab production during fatal or self-limiting malarias. Questions on Ab affinity, immunoglobulin (Ig) isotype expression, B cell induction signals, influence of host genotype on Ab-specificity, and the role of Ag-presenting B cells remain unanswered. Most of our available information has been obtained by studying inbred strains of mice with the hopes that the human situation is similar.

Abs may not only have a role in protective immunity, but also in pathogenesis. It is beyond the limits of this brief review to cover this area in more than a precursory manner.

Thus, like the disease itself, the role of Abs in malarial immunity is complex and our information fragmented. This chapter will concentrate on information obtained from the most common rodent models (mice infected with *P. berghei*, *P. yoelii*, and *P. chabaudi* and rats infected with *P. berghei*) and compare results obtained with humoral responses produced by *P. falciparum* infection in man.

II. ROLE OF ANTIBODIES IN IMMUNITY TO THE ERYTHROCYTIC STAGES

A. EVIDENCE THAT ABS ARE IMPORTANT

The ability of sera from "immune" individuals to modify the course of malaria after passive administration to nonimmune individuals is well documented.[1] In 1937, Coggeshall and Kumm[1] transferred sera from rhesus monkeys with chronic *P. knowlesi* malaria to naive monkeys infected with the homologous strain of parasite. The recipients developed malaria but survived the lethal infection. Massive amounts of sera were required to produce an inhibitory effect. To boost Ab titers, chronically infected rhesus monkeys were infected several times.[2] The resulting sera proved more effective in that a monkey receiving 10 cc of "hyperimmune" serum daily for 10 d failed to develop *P. knowlesi* malaria, and each monkey receiving 5 and 2.5 cc of hyperimmune serum daily for 10 d developed malaria but survived the infection.[2] These studies demonstrated that immunity to malaria was not totally cellular (as had been thought up to that time) and that a serum factor, presumed to be Ab, played a significant role in controlling the parasitemia.

Cohen and colleagues[3] clearly demonstrated that the effective component in sera was Abs. They purified Igs from sera of apparently immune adults residing in Gambia, an area with hyperendemic malaria, and used them to treat 12 children (aged 4 months to 2.5 years) with acute malaria. Each child received a total of 1.2 to 2.5 gm of Ig (about 10 to 20% of the child's own Ig level) at 8 to 24 h intervals over a 3 day period. Seven other children received 0.8 to 1.4 gm of the same sera but depleted in Ig. By 4 d after the initiation of treatment, parasite counts were lower in both groups, but were considerably lower (≤ 10 vs. 1000 per cubic millimeter) in children receiving immune Ig compared to Ig-depleted control sera. Immune-Ig was effective against both *P. falciparum* and *P. vivax* infections.

In the 1960s and 1970s, a variety of passive Ab studies were conducted using rodents, with *P. berghei* infection in rats and mice being the most extensively studied animal model. Diggs and Oster[4] produced hyperimmune anti-*P. berghei* sera by reinfecting rats that had survived primary infection at weekly intervals for 2 to 3 months. Young rats were injected with 0.5 ml/100 gm body weight of hyperimmune or normal rat serum and simultaneously challenged with 10^8 *P. berghei* infected-erythrocytes. All hyperimmune serum-treated rats survived infection compared to 100% mortality in controls. They also developed significantly lowered peak parasitemias (5 vs. 60%).[4] The protective activity of the hyperimmune serum was removed by adsorption with anti-Ig.

Further studies by numerous investigators used the method of passive Ab transfer to more specifically define the kinetics of protective Ab production, to evaluate ways for increasing the level of protective Ab formation, to establish the class(isotype) of protective Abs, and to further characterize protective humoral immunity.[5-11] The majority of these studies examined blood-induced *P. berghei* infection in rats or mice. Results showed that, in general, the passive transfer of Abs to rats at the same time of blood-induced (syringe passage) *P. berghei* infection resulted in a decreased onset in detectable parasitemia, reduced peak parasitemias, and enhanced survival.[5-9] The transfer of equivalent hyperimmune sera to *P. berghei*-infected mice resulted in delayed parasitemia, but the animals generally succumbed to the disease.[8,10,11] Recipients of hyperimmune sera almost always developed circulating parasitemias. Infection was only prevented in a few studies, usually by using very large doses of hyperimmune or by finely titrating the amount of hyperimmune sera transferred with the number of parasites injected.[5-8] Animals completely protected from malaria by passively transferred Ab were susceptible to infection upon rechallenge.[5] No significant difference was observed if Abs were injected immediately before, or shortly after challenge with blood-stage parasites, or if infected erythrocytes were preincubated in hyperimmune sera and then injected along with the sera.[5] Usually the transfer of hyperimmune sera (collected from animals that had been infected several times) was considerably better than sera from animals with an ongoing or chronic infection[6,10] or sera from an animal that had recovered from a single infection (immune sera).[4,5] There was generally a correlation between total anti-*P. berghei* Ab titer and protection.[5,7,8,10] The passive administration of hyperimmune sera was effective in T cell-deficient rats suggesting Abs alone were sufficient to provide protection.[9] Abs of the IgG class were found to be protective.[5,6] The protective potential of IgM Abs is difficult to assess because of their short half-life, but results for[9] and against[4,5] their protective value have been reported. In summary, results from the rat/

mouse-*P. berghei* model show that passive transfer of antimalarial IgG in sufficient quantitites can delay the onset of infection, decrease peak parasitemias, and enhance the survival of infected rats.

Abs also are effective in the control and elimination of the erythrocytic stages of *P. yoelii* in mice. *P. yoelii* generally produces a self-limiting infection in mice. BALB/c mice receiving 0.5 ml of hyperimmune sera at the time of challenge with 10^4 infected erythrocytes failed to develop patent infections[12] and those receiving 0.3 ml had significantly delayed prepatent periods (14 d compared to 5 d in the controls).[12] Even passive transfer of sera from convalescing mice at the time of challenge delayed the course of infection in recipient animals.[12] The transfer of hyperimmune sera to BALB/c or outbred mice with ongoing acute *P. yoelii* infections resulted in a rapid decrease in parasitemia to <0.1% within 2 d.[12,13] Maximal levels of protective Abs were produced after a secondary challenge. It was suggested that Abs alone were sufficient to control *P. yoelii* parasite replication since the initial transfer of 0.5 ml of hyperimmune or several injections of convalescent sera prevented mice from developing patent *P. yoelii* infections, and there was not evidence that the mice themselves had responded immunologically.[12] However, other investigators found that doses of serum that completely protected intact mice from infections were insufficient when administered to T cell-deprived mice, suggesting that the induction of a host T cell response was required to prevent infection.[14]

P. chabaudi also produces a self-limiting infection in many strains of mice (see Chapter 7). It has been shown that the passive transfer of small amounts of immune (sera collected following a single infection) or hyperimmune (sera collected after two or more infections) at the time of challenge results in a delay in patency and reduces peak parasitemias.[15] However, the administration of large amounts of high titered hyperimmune serum at the time of challenge exacerbates the infection.[16] The negative effect of passive transfer of hyperimmune serum has been reported from other malarial host-parasite systems.[7] Thus, in the *P. chabaudi* model, the protective activity does not always correlate well with Ab titer.[15-16]

The above studies show that the passive transfer of anti-malarial Abs can help eliminate or prevent blood-stage parasitemias. It seems logical, therefore, that if animals were deficient in B cells they would develop more severe malarial infections. Weinbaum and colleagues[17] and Roberts and Weidanz[18] investigated the importance of B cells in immunity to *P. yoelii* using μ-suppressed mice (Figure 1). BALB/c mice, injected three times a week with anti-μ antisera from birth, are functionally B cell deficient at ~8 weeks of age, but have relatively normal T cell functions including the ability to reject foreign skin grafts and proliferate *in vitro* in response to PHA, Con A, and allogeneic cells.[17] When μ-suppressed mice are injected with 10^5 *P. yoelii* (17XNL avirulent) parasites, they develop rapid infections with parasitemias reaching >35% by day 7 (compared to 7% in intact controls) and 100% mortality by day 22 (Figure 1). Thus, without Abs, BALB/c mice are not able to control a normally self-limiting primary *P. yoelii* infection. To determine if immunity to *P. yoelii* could be established in the absence of B cells, 11 B cell-deficient BALB/c mice were infected with *P. yoelii* and then treated with the drug clindamycin to prevent death.[18] Then seven of the mice were challenged with *P. yoelii*. Six of the mice developed chronic plasmodial infection; the remaining mouse died. Of the six mice with chronic infections, three produced antimalarial Abs (i.e., broke μ-suppression), but the other three maintained long-term chronic infections in the apparent absence of Abs.[19] This single study, which used μ-suppressed mice and drug rescue, suggests that BALB/c mice are able to develop an alternative, Ab-independent mechanism to survive *P. yoelii* infections.

In contrast to *P. yoelii,* mice appear to easily control primary *P. chabaudi adami* infections in the total absence of Abs.[19] Grun and Weidanz (1981) infected eight B cell-deficient and seven normal (C57BL/6 × CBA)$_{F1}$ mice with 10^5 *P. chabaudi adami*-infected

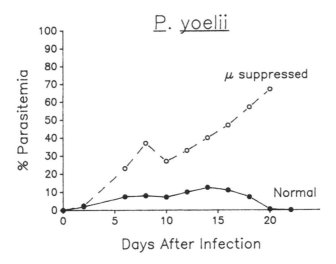

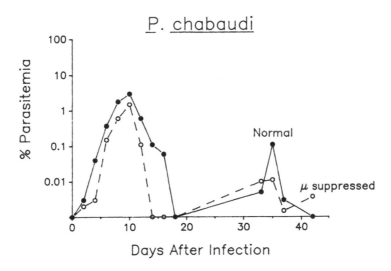

FIGURE 1. Course of *P. yoelii* and *P. chabaudi* infection in normal and μ-suppressed BALB/c mice. Mice were infected with 10^5 parasitized erythrocytes on day 0 and percent parasitemia (number infected erythrocytes/total erythrocytes) was determined. Adapted from References 17 and 19.

erythrocytes and found that the course parasitemia was identical between the two groups (Figure 1). The B cell-deficient mice, however, were unable to ultimately eliminate the parasites, whereas normal mice did. *P. chabaudi*-infected mice can apparently control acute infection in the absence of Abs, but require Abs to eventually clear the infection. Clearly, the importance of Abs in controlling primary acute *P. chabaudi* and *P. yoelii* infections differs. This emphasizes the statement made in the introduction that one must be careful in making generalizations about humoral immunity when studying different species of parasites.

Immunity to rodent malarial parasites can be adaptively transferred with cells as well as serum.[20-26] Early studies showed that the injection of lymphocytes (generally splenocytes) from immune donors to naive, syngeneic recipients resulted in decreased parasitemias in rats infected with *P. berghei*[20-22] and in mice with *P. berghei*,[22] *P. yoelii*,[23,24] and *P. chabaudi*.[25,26] Several researchers have investigated which lymphocyte subset(s) was responsible for transferring protection. The greatest protection was achieved using immune B cells.[22-25] Although some protection could be demonstrated in some systems by transferring immune T cells, the authors concluded that the T cells probably mediated their effect by serving as helper cells for Ab production.[23,26]

In summary, in every animal model studied to date, including man, the passive transfer of antimalarial Abs results in some diminution of parasitemia, often leading to survival of the host. Partial immunity against rodent malarial parasites including *P. berghei, P. yoelii,* and *P. chabaudi* can be adaptively transferred to naive mice or rats via immune B cells. In the absence of B cells, mice die from an otherwise nonlethal *P. yoelii* infection. Even in the *P. chabaudi* model, where mice can modulate acute *P. chabaudi* infection in the absence of Abs, Abs are required for eventual elimination of the parasites. Thus, based on available data, Abs appear to constitute an important defense mechanism for the host against malarial parasites.

B. ASSESSING "PROTECTIVE" ANTIBODIES

Many investigators have tried to develop serologic assays that correlate with protective immunity. In 1969, Diggs and Oster[4] used probit analysis to quantitate the amount of protective Abs present in sera. They defined protective Abs as the highest dilution of sera that, when injected into mice with a standard number of *P. berghei*-infected erythrocytes, suppressed parasitemias by 50% 3 d later. They followed the kinetics of protective Ab formation in *P. berghei*-infected rats. Results of that study are reproduced in Figure 2. Rats cleared peripheral parasitemias about 24 d after infection. At that time, protective Ab could be detected in their sera. Over the following 7 months, titers remained virtually the same, except that higher levels of protective Ab (comparable to those seen in hyperimmune sera) were observed about 3 months after infection. Therefore, protective Abs (as defined as Abs capable of decreasing parasitemia upon passive transfer) were detected following recovery with levels being elevated immediately after clearing the infection. To date, similar quantitative studies in other animal models have not been reported.

In humans exposed to malaria, several years may be required before immunity is achieved.[27,28] The rate of establishing maximal levels of protective Abs may correlate with endemicity of the parasites. In rodents, the situation appears somewhat different as maximal Ab titers and the production of "protective" Abs can frequently be detected after two exposures to the parasite (i.e., recovery from primary infection followed by a second infection).

It is not clear why several exposures are required to produce significant levels of protective Abs. Repeated exposure could result in (1) increased Ab titers, (2) recognition of Ags not previously recognized (i.e., minor antigenic determinants or antigenic variation), and/or (3) increased Ab quality (i.e., affinity, avidity, isotype, etc.). Undoubtedly, all three are important.

To help clarify which of the above considerations might be important, we used the mouse malaria system. BALB/c mice were infected with 17XNL *P. yoelii,* and serum samples were collected 2 weeks after infection (acute sera), 2 weeks after clearing the parasites (day 32), and 2 weeks following two to eight challenges with the lethal strain of *P. yoelii.* Sera were tested for (1) Ab titers (total and isotype specific) by radioimmunoassay (RIA), (2) ability to passively transfer protection, and (3) used to immune precipitation ^{35}S-methionine-labeled *P. yoelii* Ags. Results are summarized in Table 1 and Figure 3. As expected, differences in sera collected during and following primary and after secondary exposure were found. Ab titers increased between primary and secondary infection (from 1:100,000 to 1:1,000,000), the increase being due to elevated IgG$_2$ and IgG$_3$ titers (Table 1). Sera collected after primary infection modulated the infection on passive transfer, but not as significantly as sera after secondary infection (Figure 3). Sera collected following two to eight malarial exposures had essentially equivalent high Ab titers (Table 1), and were equally efficient in transferring protection. Immune precipitation studies did not reveal the presence

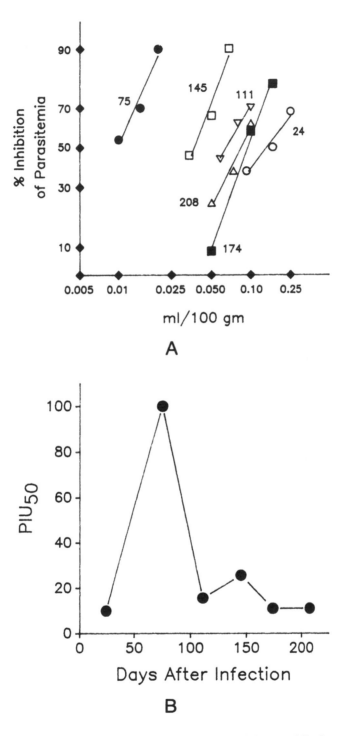

FIGURE 2. Quantitation of protective antibody levels in serum following *P. berghei* infection in rats. Sera were collected from six groups of rats at various times after primary *P. berghei* infection. Three doses of each convalescent sera were passively transferred along with a standard number of parasitized erythrocytes into rats, and percent inhibition of parasitemia for each dilution is plotted in A. The numbers appearing on the figure refer to the day the serum was collected. Data were converted into 50% parasite-inhibiting units (PIU_{50}) by dividing one by the amount of serum (ml/100 gm) producing 50% reduction in parasitemia. PIU_{50}/ml of serum is shown as a function of time after infection in B.

TABLE 1
Anti-*P. yoelii* Ab Titers in Pooled Plasma Samples Collected During Primary Infection and Following Repeated Exposure

	IFA	Log_{10} RIA Titers				
		Total[a]	IgM	IgG_1	IgG_2	IgG_3
During self-limiting						
Plasmodial yoelii						
Day 6		1.0	<1.0	<1.0	1.1	1.1
Day 7	1:16	2.0	1.2	1.2	3.0	2.1
Day 8	1:64	2.0	1.0	<1.0	2.0	2.2
Day 9	1:64	2.1	3.1	<1.0	2.0	2.1
Day 14	1:128	3.1	5.3	2.1	2.6	3.2
Following primary infection						
Day 32	1:2,048	5.1	4.1	4.1	5.1	4.2
Following reinfection						
2°	1:16,384	6.2	4.1	4.2	6.2	5.1
3°	1:4,096	6.3	3.2	4.1	6.1	4.2
6°	1:4,096	6.4	4.2	5.2	6.3	4.1

[a]   RIA using ^{125}I-GaMIg (polyvalent).

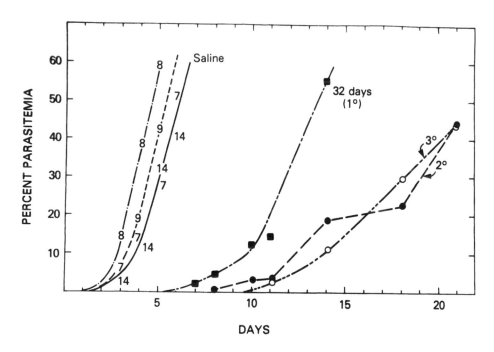

FIGURE 3.   Course of 17XL *P. yoelii* following passive transfer of sera collected during infection following primary infection, and repeated exposure. Groups of BALB/c mice were injected with 0.5 ml of serum 4 h before injection of 10^6 17XL *P. yoelii* (lethal) parasites. The sera are characterized in Table 1. Sera collected on days 7, 8, 9, and 14 of infection did not alter the course of infection. Although sera collected after clearance of primary parasitemias (32 d 1°C) and following two to eight reinfections significantly delayed patency and prolonged death, all mice ultimately died.

of Abs following secondary or multiple infection with antigenic specificities (which might correlate with protective Abs) that were not present following primary infection. The ability to alter the course of infection on passive transfer correlated more with Ab quality than any other parameter in this study.

With the exception of the quantitative passive Ab transfer methodology described above,

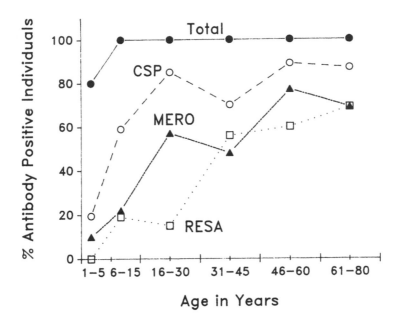

FIGURE 4.   Proportion of individuals with antibodies to 3 well-defined malarial antigens. Sera from 144 individuals living in an area with mesoendemic *P. falciparum* malaria was tested (1) by IFA for anti-*P. falciparum* Abs, (2) by ELISA for Abs to portions of the circumsporozoite protein (CSP) and the merozoite-surface antigen Pf195 (MERO), and (3) by IFA on glutaraldehyde-fixed smears of *P. falciparum* for RESA. See text for details. (Adapted from Chizzolini, C. A. et al., *Am. J. Trop. Med. Hyg.*, in press).

there is no other way to demonstrate the presence of protective Abs. Although individuals with high anti-malarial Ab titers are more likely to be immune than those that do not, there is no strict correlation between Ab titers and *in vivo* immunity. Sera from immune individuals can alter *P. falciparum* growth *in vivo*,[29-32] but not all immune individuals have inhibitory Abs and not all individuals with inhibitory activity are immune (see Section C).

The most recent approach for identifying protective Abs serologically involves using genetically engineered or biochemically synthesized malarial Ags. Sera collected in endemic areas from individuals of known ages are tested for Abs against individual peptides. It is hoped that high Ab titers against one or several Ags will correlate with immunity. A summary of a recently completed study by Chizzolini and colleagues,[33] conducted in an area with mesoendemic *P. falciparum* malaria in Gabon, West Africa, is shown in Figure 4. Ab titers to (1) a polypeptide consisting of 40 (Asn-Ala-Asn-Pro) repeats of the *P. falciparum* circumsporozoite protein (NANP), (2) the fusion peptide 31.1 consisting of the N-terminal portion of the 190 to 200-kDa glycoprotein found on the surface of merozoites, and (3) the ring-infected erythrocyte-surface antigen (RESA, Pf155) were determined.[33] Results show that about 90% of the individuals made Abs to the CSP protein and that maximal, stable adult Ab levels were reached between 10 to 30 years of age. On the other hand, only 70 to 80% of the population had Abs to the merozoite and erythrocyte-surface Ag, and stable levels were not reached until ~31 to 45 years of age (Figure 4). Comparison of Ab titers between infected and uninfected individuals suggested that the Ab response to the blood-stage Ags correlated more with resistance against malaria in subjects in the Kassa district, better than Ab titers to prehepatic stage Ags of *P. falciparum*. Similar results have been reported by others. Further studies are needed to determine if this approach will lead to the development of a serologic assay for detecting protective Abs.

C. POTENTIAL EFFECTOR FUNCTIONS OF "PROTECTIVE" Abs

Several review articles have discussed potential effector functions of Abs,[34,35] so only a brief update will be presented here. There are numerous, obvious mechanisms by which Abs could aid in the elimination of malarial parasites, but the relative importance of each of these methods remains unresolved. Abs to antigens located on the surface of cells, either the parasite itself[30-32,36-51] or on infected erythrocytes,[52-61] are held to be the most important.

Numerous studies have clearly demonstrated the effectiveness of Abs in controlling *P. falciparum* and *P. knowlesi in vitro.*[30,31] IgG Abs are more effective than IgM, and inhibitory Abs require at least two binding sites per molecule.[30] There is little or no evidence that Abs influence the development of parasites within the erythrocyte.[30-32,37] Alternatively, Abs are thought to bind to merozoite surface Ags, resulting in agglutination of extracellular merozoites or blocking of surface receptors involved in penetration of erythrocytes.[36,38,39] Monoclonal Abs (mAbs) to antigens located on merozoites of *P. falciparum* (Pf195)[40] and *P. knowlesi* (Pk140)[41] affect parasites *in vitro* and mAbs to *P. yoelii* (Py230)[42,43] and *P. chabaudi* (Pc250)[44] are effective *in vivo*. Abs to other merozoite Ags undoubtedly are present in immune sera.

Unfortunately, there is imperfect correlation between the presence of Abs that are inhibitory *in vitro* and clinical immunity. This was particularly well documented in the study of Fandeur et al.[45] using *P. falciparum*-infected squirrel monkeys *(Saimiri sciureus)*. Sera from some immune monkeys could adaptively transfer immunity to naive recipients (estimated serum concentration 0.1 mg/ml), but some were ineffective at blocking parasite invasion *in vitro* when present a 1 mg/ml. Other sera were effective *in vitro*, but not *in vivo*. Purified IgG from "immune" sera transferred protection, but F(ab)$_2$ preparations thereof were ineffective. These results suggest that *in vitro* inhibition assays primarily measure Abs directed against merozoites or erythrocyte-expressed surface Ags, (see below), but *in vivo* protective Abs are active through a more complex immune mechanism, probably involving immune phagocytosis and possibly other cell-mediated immune mechanisms.

Abs are known to enhance immune (Fc) phagocytosis of parasites.[46-51] This is particularly effective when freed parasites or damaged, infected erythrocytes are studied.[49,50] Immune phagocytosis is clearly an important mechanism *in vivo* for removal of dead parasites, but its true role in parasite killing and elimination remains unclear. However, recently, Druilhe and Khusmith[51] used a phagocytosis assay to measure cytophilic antibodies against *P. falciparum* merozoite surface antigens. They found that Abs that mediated phagocytosis of merozoites increased slowly and gradually with age in humans, and that production of high titers required about 15 years of continuous exposure to malaria. The appearance of antibodies promoting merozoite phagocytosis concurred with the ability of exposed individuals to control high parasitemias as well as the pathologic consequences of infection, such as spleen enlargement. There was no direct correlation between total Ab titers and phagocytic enhancing Abs. The authors suggest that the appearance of Abs that mediate phagocytosis of merozoites more closely correlated with protective immunity than any other known parameter.[51]

Antigens expressed on the surface of infected erythrocytes are certainly targets for Ab-mediated interactions. However, there is no solid evidence that infected erythrocytes can be lysed by Ab and complement.[52] Abs to malarial Abs on the erythrocyte surface may mediate immune phagocytosis (intact erythrocytes are occasionally seen within macrophages),[52] and prevent cytoadherence.

The most well-characterized Ags of infected erythrocytes are the knob-associated Ags (K$^+$) of *P. falciparum*.[52-58] Electron-dense protuberances appear on the erythrocyte surface and mediate cytoadherence of infected erythrocytes to endothelial cells in the deep vasculature[52,53] and contribute to cerebral malaria.[54] Recent studies show that the histidine-rich, knob-associated protein is not directly responsible for cytoadherence;[55,56] thus, the cytoadherent surface ligand remains to be defined. To test the effects of Abs on cytoadherence, Schmidt et al.[57] and Udeinya et al.[58] developed an *in vitro* assay for the binding of infected erythrocytes to amelanotic melanoma cells. They reported that Abs in immune sera prevent or reverse the binding of K$^+$ *P. falciparum*-infected erythrocytes to melanoma cells in a strain-specific manner.[58] David et al.[59] demonstrated that injection of immune sera with high titers of "cytoadherence blocking Ab" *in vitro* resulted in the release of sequestered erythrocytes in *P. falciparum*-infected squirrel monkeys.[59] Thus, Abs to erythrocyte surface Ags are important in preventing sequestration *in vivo*.

A second *P. falciparum* Ag, known to be associated with the infected erythrocyte membrane is Pf155, or RESA.[60,61] This Ag is inserted into the erythrocyte membrane during merozoite invasion. Abs to this Ag alter parasite growth *in vitro*,[60] and *Aotus* monkeys vaccinated with a peptide segment of the antigen were partially protected from *P. falciparum* challenge.[61] Currently, it is unclear how Ags to Pf155 actually mediate their protective effects.

In summary, Abs to cell-surface merozoite and erythrocyte Ags are known to mediate parasite elimination. Certainly, Abs are important in the phagocytosis of both intact and soluble parasite Ags. Other proposed mechanisms, including complement-mediated parasite lysis, antibody-dependent cytotoxicity (ADCC), blocking of merozoite "penetration receptors", blocking of nutrient flow into or waste product removal from the erythrocyte, blocking the parasite transferrin receptor, etc., have not been firmly established.

D. T CELLS ARE REQUIRED FOR Ab PRODUCTION

It is well documented that T cells are important in malarial immunity. This was shown in early studies, in which mice were rendered T cell depleted (deprived) by treatment with anti-lymphocyte serum,[62] adult thymectomized, irradiated and bone marrow reconstituted,[14,25,63,64] or were congenitally athymic (nu/nu mice).[17,65] T cell-depleted mice infected with lethal *P. berghei*,[62] nonlethal *P. yoelii*,[14,63-65] and nonlethal *P. chabaudi*[19,25] developed fulminating parasitemias following blood-induced challenge. In *P. yoelii* infections, the course of parasitemia was identical in normal and T cell-deprived mice until day 10 to 15 of infection, such that the effects of T cells are not seen until reasonably late in infection.[17] Jayawardena et al.[64] reported that anti-*P. yoelii* Ab titers were significantly reduced in T cell-depleted (thymectomized, irradiated, bone marrow reconstituted) mice, especially of the IgG$_1$ class. Reconstitution of T cell-deficient mice with normal or immune T cells restored immunity and Ab production.[64]

Jayawardena et al.[63] showed that reconstitution of T cell-deficient mice with CD4$^+$ (ly1$^+$2,3$^-$, L3T4$^+$) cells, but not CD8$^+$ (ly1$^-$,2,3$^+$) cells restore immunity to *P. yoelii*. The conclusion drawn was that the major role for T cells in *P. yoelii* infection was to provide help for Ab production. Brinkmann et al.[66] also reported that ly2$^-$ cells were the most effective in transferring immunity, but found that ly2$^-$ cells were insufficient to completely protect C57B2/6 nu/nu mice. It was thought that contact with other T cells or their lymphokines, which would be present in irradiated mice, were required. The requirement for cells, other than ly2$^-$ T cells and B cells, in humoral immunity cannot be assessed since Ab titers were not evaluated in this study. It has recently been shown that immunity to *P. yoelii* can be produced in T cell-deficient (thymectomized, irradiated, bone marrow reconstituted) CBA mice by the adaptive transfer of CD8$^+$ (Lyt1$^-$ly1$^-$2,3$^+$) cells as long as an IJ$^+$,Ig$^-$,thy1$^-$ cell is also transferred.[67] Unfortunately, Ab titers were not measured in these animals, but one would assume that anti-malarial Ab titers would be low due to the absence of T$_H$ cells. These results suggest that there may be a role for T cells in *P. yoelii* immunity in addition to providing help for Ab induction.

In *P. chabaudi* infections, T cell-depleted mice developed acute infections and rapidly died without any indication that they had even partially controlled the infection.[19] Immunity against this parasite can be transferred with CD4$^+$ (ly1$^+$2,3$^-$) lymphocytes.[19] Because the

T-cell effect is so rapid, it is thought the T cells may function in protective mechanisms, in addition to helping Ab production. These studies are discussed in detail elsewhere in this book.

Significantly depressed antimalarial Ab titers were found by Jayawardena et al.[64] in T cell-deficient mice, and Weinbaum et al.[17] reported antimalarial titers of only 1:4 (by passive hemagglutination) in BALB/c nu/nu mice. We sought to further define the requirement for T_H cells for antimalarial Ab production, and reasoned that if Abs were produced in the absence of T cells they would be induced by type 1 or 2 T-independent (T-I) Ags.[65] Previous studies had reported finding a lipopolysaccharide (LPS)-like substance in *P. falciparum*. BALB/c nu/nu and nu/+ littermates were infected with 17XNL *P. yoelii*, and antimalarial Ab activity was determined throughout the infection using an isotype-specific RIA assay. Antimalarial Abs were detected in nu/+ but not nu/nu mice using this very sensitive assay, highly suggesting that *P. yoelii* parasites do not possess TI Ags. Since nu/nu mice are genetically deficient, it is possible that they had an altered B cell repertoire. To test this, lymphocytes from *P. yoelii*- infected nu/nu mice were used in lymphocyte hybridization studies. Anti-*P. yoelii* hybridomas secreting IgM monoclonal Abs were produced, showing that nu/nu mice possess antimalarial B cells that become activated during malarial infection, but T cell signals (probably both direct contact and lymphokine differentiation signals) are required to commence isotype-switching and Ig secretion.

E. LITTLE IS KNOWN ABOUT MACROPHAGES AS ANTIGEN PRESENTING CELLS (APC)

Undoubtedly, macrophages and related RES cells are important in the processing and presentation of malarial antigens for T-cell activation. At least one study has shown the preferential binding of immune lymphocytes to macrophages presenting plasmodial Ags *in vitro*.[68] Recently, studies using peptides of the circumsporozoite protein (CSP) antigen have begun to detail antigen presentation (see Section on sporozoites below), but similar studies on the kinetics, size, and nature of erythrocytic stage antigens presented by macrophages have not been published to date.

It is known that during acute malarial infections defects in macrophage functions occur (see Chapter 4 for details). Some of these defects affect humoral immunity. Brown and colleagues[69] clearly demonstrated that splenic, and to a lesser extent, peritoneal macrophages were defective in mediating Ab production to the T cell-dependent antigen, sheep red blood cells (SRBC), but not to the T-I antigen, TNP-ficoll. In these studies, splenocytes from mice acutely infected with *P. berghei* (day 8 of infection) and normal mice were removed and cultured *in vitro* with SRBC. The number of plaque forming cells (pfc) was determined 4 d later. Only 900 pfc per culture were found using cells from infected mice compared to 2000 pfc per culture for normal mice (a 220% reduction). To determine if the reduction was due to adherent (macrophages) or nonadherent (lymphocytes) cells, adherent cells from infected mice were cultured with nonadherent cells from normal mice. Only 270 pfc per culture were observed. Conversely, culturing nonadherent cells from infected mice with normal splenic or peritoneal macrophages resulted in 1600 and 2000 pfc per culture, respectively. These results show that the defective component in the humoral immune response was in the adherent cell population and not in the lymphocytes. PFC responses to the TI-antigen TNP-ficoll were not affected by malarial infection. The defect in macrophages could be due to a defect in "antigen handling" or monokine production.[70] The former is the more likely explanation since monokine production is important in TI-responses and this response was not affected. Additionally, during acute *P. yoelii* infection there is no evidence for altered Interleukin-1 (IL-1) levels, though IL-2 levels are decreased.[71]

Biozzi and colleagues[72] produced low (L) and high (H) Ab responding strains of mice. L responder mice produced low Ab titers when immunized with *P. berghei* compared to H responder mice. It was shown that *P. berghei* antigens were significantly more rapidly destroyed in L responder animals.[72,73] Thus, that rate of antigen catabolism strongly influences the level of Ab production. Clearly, studies on antigen processing, presentation, and catabolism by normal and activated macrophages are needed before humoral immunity to erythrocytic stages antigens can be understood.

F. B CELL SUBSETS INVOLVED IN HUMORAL IMMUNITY

There are at least two B-cell subsets in mice that can be distinguished by the lyb5 phenotype.[74] Lyb5[+] cells respond to both T-I- and T-dependent antigens, are selectively stimulated by high concentrations of Ag, and secrete predominantly IgM antibodies.[74] The other subset, lyb5[-], does not respond to T-I Ags, requires T cell help (possibly through direct cell contact), is selectively stimulated by low concentration of Ag, and frequently secretes IgG Abs. Lyb5[+] cells do not need IL-4 or IL-5, whereas lyb5[-] cells do, for growth and differentiation. In most mouse strains, an equal number of lyb5[+] and lyb5[-] B cells are present. However, CBA/N mice lack the lyb5[+] subset of B cells.

In 1979, Hunter et al.[75] and Jayawardena et al.[76,77] compared the course of *P. yoelii* infection in male (BALB/c × CBA/N)_{F1} mice that are immunologically normal and male (CBA/N × BALB/c)_{F1} mice that lack lyb5[+] B cells (i.e., B cell defective). Almost identical results were obtained by the two groups. The course of *P. yoelii* infection was similar in B cell normal and defective mice until days 8 to 10 of infection. After that, defective mice developed 2 to three 3 times higher percent peak parasitemias that persisted for a longer period (24 to 26 d compared to 17 to 18 d in normal controls), and some mice died. Thus, the absence of lyb5[+] B cells greatly hampered control of infection.

In both studies, a severe depression of the antiparasite IgM response was seen. On day 11 of infection, Jayawardena et al.[77] reported IgM IFA titers of only 1:8 in defective animals but compared to 1:512 in normal controls, and on day 12, Hunter et al.[75] reported RIA IgM titers of only 1:2.5 in defective compared to 1:77 in normal controls. Initially, the IgG response was slow, but was equal by day 21 of infection of both groups of animals. Thus, CBA/N mice developed more severe courses of infection and depressed IgM responses, but IgG responses were relatively normal.

Jayawardena et al.[71] further evaluated the response of Lyb5[+] B cells in the absence of T cells. CBA mice were thymectomized at 10 weeks of age, irradiated with 1000 rads, and then reconstituted with bone marrow from normal or B cell-defective mice. Following challenge, both groups produced minimal levels of antimalarial Abs, and ultimately died of "nonlethal" malaria. This study demonstrates that neither lyb5[-] or lyb5[+] + lyb5[-] cells alone are sufficient to mediate immunity in the absence of T cells. Adaptive transfer studies were conducted to validate that the defect was in the B cell compartment only. B cells from normal immune mice could adaptively transfer partial protection. This effect was enhanced by the simultaneous transfer of immune T cells. Transfer of immune B cells from normal mice plus immune T cells from B-cell defective mice also provided partial protection. However, protection could not be transferred by immune B cells from B cell-defective mice plus normal immune T cells. Thus the defect was in the B not the T cell compartment.

Once having recovered from *P. yoelii* infection, normal and defective mice are equally efficient at handling secondary *P. yoelii* challenge. The enigma remains that although CBA/N mice become immune, their immune lyb5[+] B cells cannot adaptively transfer protection. This suggests that primed T cell help may be required for immunity induced by both lyb5[+] and lyb5[-] cells.

These studies by Hunter, Jayawardena, and colleagues demonstrate that both B cell populations are involved in inducing immunity, that the IgM response is primarily under T cell regulation, and that both IgM and IgG antimalarial Abs may be important in mediating maximal immunity. Equivalent B cell subsets have not been found in humans.

G. KINETICS OF Ab PRODUCTION AT THE SERUM AND CELLULAR LEVEL

The rate of appearance of antimalarial Abs appearing in sera following primary infection (usually blood induced) have been studied in various malarial rodent, primate and human host-parasite systems. Figure 5 provides a comparison of the primary Ab response in mice and man following blood-induced *P. yoelii*[65] and *P. vivax*[78-80] infection, respectively. Results are compared with these for sporozoite-induced *P. vivax* infections. In general, within 2 to 3 d after detecting parasites in the peripheral circulation, antimalarial Abs can be detected by indirect immunofluorescence (IFA). This is true for both blood-induced and sporozoite-induced infections. Ab titers increase rapidly, reaching IFA titers of 1:1280 to 1:5120 within 15 d in *P. yoelii*-infected mice and 5 d in *P. vivax*-infected humans. The rate of increase appears more rapid in humans than mice, which may not be too surprising as mice develop high parasitemias. Ab titers gradually decline, but antimalarial Abs can easily be detected 3 months after infection (see Section K). The total Ig response, including hypergamma-globulinemia, parallels the Ag-specific response (see below). Similar results have been reported for *P. falciparum* infection in man.[81]

Information on Ab production at the B cell level is very limited. In rodent models, combined results from several studies show that antimalarial Ab production is initiated during the 1st week of infection. We have collected splenocytes on day 4 following blood-induced 17XL *P. yoelii* infection and successfully produced hybridomas secreting IgM anti-*P. yoelii* antibodies. These results demonstrate that resting B cells had become activated and were undergoing clonal expansion at day 4. Results of Jayawardena et al.[64] support early Ab production. On day 7 of *P. yoelii* infection, they found that pyroninophilic lymphoid cells were accumulating in the thymus-dependent periarteriolar regions of the spleen and that small peripherally located germinal centers were present. Blast cells, dividing cells, mac-rophages, and cell debris were present within germinal centers. By day 14 to 21, splenic germinal centers were large and active. As early as day 5 to 6, anti-*P. yoelii* Abs can be detected in the peripheral blood of CBA, C₃H, and BALB/c mice.[64,65,82] Thus, the initial kinetics of anti-*P. yoelii* Ab production is similar to that following exposure to other well-studied antigens.

As would be predicted for a bloodborne disease, the spleen appears to be the primary source of Ab production. Lymphocyte activation and division, increased numbers of plasma cells, and germinal center formation are evident in the spleen days 7 to > 28 in *P. yoelii*-infected mice.[64] No changes were observed in axillary and mesenteric lymph nodes, Peyer's patches, or the thymus of infected mice.

Little is known about the number of malarial Ag-reactive B cells present in the spleen of naive animals or the number of memory B cells following recovery. The reason for this is because it has been extremely difficult to develop an Ag-specific antimalarial pfc assay. Recently, Wassom et al.[83] developed a filter immunoplaque assay for quantitation of parasite-specific antibody producing cells. In this assay, soluble malarial antigens, attached to ni-trocellulose membranes, are overlaid with a monolayer of splenocytes from infected mice. Abs secreted by individual plasma cells bind to nearby Ags on the membranes, which are subsequently treated with enzyme-labeled antiimmunoglobulin and substrate. In preliminary studies, they reported that on day 8 of *P. yoelii* infection, >60 to 100 cells per 10⁶ splenocytes were secreting IgM anti-*P. yoelli* Abs (0.02% of B cells), whereas only ~7/10⁶ splenocytes were secreting IgG Abs.[83] Similar results were observed on day 15 of infection. Undoubtedly, these results are on the low side since only soluble malarial Ags were used, and many immunogens are associated with malarial membranes. The range of reactivity, however, is comparable to that reported for the number (0.02%) of B cells responding to a simple Ag in naive mice.[84] The assay shows promise for providing important information on the anti-malarial B cell repertoire in mice.

Wen et al.[85] reported using a limiting dilution assay for enumerating *P. falciparum*-reactive B cells in the blood of naive humans and those with a history of malaria. In this assay, known numbers of surface Ig$^+$ human B cells are cultured with the mouse thymoma line EL4 which induces polyclonal activation, and with supernatants from human T cell and macrophage cultures which served as a source of proliferation and differentiation factors.[85] In this assay, individual B cells undergo several rounds of proliferation, producing ~20 ng of Ig. Cultures containing anti-*P. falciparum* Abs were detected by indirect IFA (sensitivity ~1.6 ng/ml). Results showed that in naive individuals, ~0.01 to 0.1% (1/3780 B cells) peripheral blood B cells could produce antimalarial Abs, whereas the level was approximately ten times higher (mean 1/320 B cells) in individuals who had had malaria (range 1/96 in an individual who had had repeat infections to 1/851 in an individual with an apparent single infection). Unfortunately, data on the length of time since last exposure to malaria were not provided, but it appears the number of circulating Ag-reactive B cells is significantly elevated for a long period following recovery.

In addition to limited information on the number of activated B cells, there is little data on the Ag specificity of responding B cells; that is, it is uncertain if, during the first few days of infection, *all* antimalarial B cells become activated or only those that respond to "immunodominant" epitopes. In the first case, Abs to all malarial epitopes (including potentially "protective" epitopes) would be present throughout the infection. In the latter case, a sequence of Abs would appear in sera during the course of infection. We attempted to address this question using the BALB/c *P. yoelii* model. The basic procedure was to collect splenocytes during the course of primary infection, wash them free of as much extraneous (free) parasite material as possible, incubate the cells *in vitro*, and then determine the Ag specificity of the newly secreted Abs by immunoprecipitating metabolically labeled *P. yoelii* Ags. Protein A was used, so IgG responses were predominantly measured. Representative results, shown in Figure 5, suggest that a series of Abs are produced to an increasing number of Ags during infection. On day 6 of *P. yoelii* infection, spleen cells were secreting Abs to at least 16 *P. yoelii*-Ags. Very few of those Abs could be detected in plasma at day 6, but all were detectable in plasma by day 13. On day 13 of infection, spleen cells were secreting Abs to at least 14 Ags, at least one of which, to an Ag of Mr $\cong$230,000, had not been present at day 6. This is particularly interesting as several studies have shown that monoclonal Abs to a 230,000 Ag, located on the surface of merozoites, are important in controlling the infection.[42,43] By day 20, when parasites had been eliminated from the peripheral blood, spleen cells continued to synthesize Abs to at least 20 to 25 immunogens. Spleen cells collected on day 27 produced Abs to a reduced number of Ags.

H. POLYCLONAL VS. AG-SPECIFIC IG SECRETION

It is well known that large amounts of immunoglobulin are produced during malaria infection in mice and man.[3,14,37,86-89] Increases in Ig levels of ~8 mg/ml have been reported in man,[78,80,86,89] 10 to 12 mg/ml in mice infected with *P. yoelii* and *P. berghei*,[17,65,88] and ~5 mg/ml in *P. chabaudi*-infected mice.[90] However, it is clear that only a small fraction of the total Ig increase has antimalarial reactivity. In 1964, Curtain et al.[86] collected plasma from 121 Melanesian children (aged 0 to >10 years) residing in areas with holoendemic malaria in New Guinea and New Britain. Plasma from these individuals was passed over columns coupled with either an extract of uninfected- or *P. falciparum*-infected erythrocytes, and the amount of γ_2-7S Ig present before and after adsorption was determined. Approximately 0.5% of Ig was retained by the RBC column, and 6 to 11% by the malaria column. Thus, an estimated 5.5 to 10.5% of the total Ig was found to be malaria-specific, suggesting that there was 1.7 to 2.9 mg/ml of anti-*P. falciparum* Abs circulating in the blood of these individuals. Similar studies by Cohen, et al.[37] using various intra- and extracellular preparations of *P. falciparum* parasite to adsorb normal (nonspecific adsorption control) and

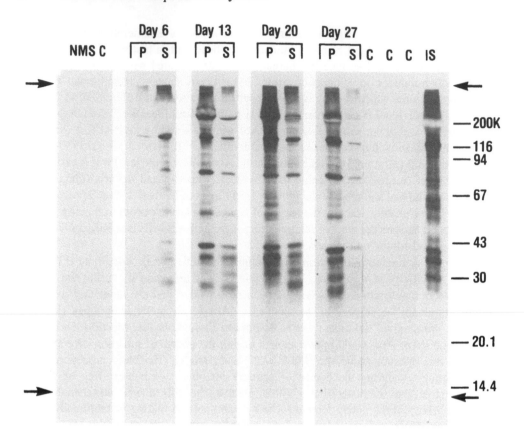

FIGURE 5.   Sequence of *P. yoelii* antigens inducing an Ab response during primary infection in BALB/c mice. Spleens were removed from mice during *P. yoelii* infection, washed free of extraneous Ag, and then cultured for 24 h *in vitro*. Azide was added to control cultures (C) to verify that the Abs were synthesized and not shed. Paired culture supernatants (S) and plasma (P) were used to immunoprecipitate ^{35}S-methionine-labeled *P. yoelii* proteins. Normal mouse serum (NMS) and immune serum (IS) were used as negative and positive controls, respectively. See text for further details.

immune sera showed that the maximal level of specific antibody was 5.4% of the total IgG in immune Gambian sera and 2.6% in New Guinea sera. A lower percentage was obtained with the *P. knowlesi*-rhesus monkey system.[30,87] IgG, purified from normal and immune monkey sera by DEAE chromatography, was labeled with [125]I − or [131]I − , and then adsorbed with a large number of *P. knowlesi* schizonts obtained by lysis of infected erythrocytes.[87] Only 0.9% of the total IgG was found to have antiparasite reactivity. In adsorption studies, it is difficult to insure that Abs against all malarial determinants (membrane bound, internal, soluble, etc.) are removed but it is clear that only a minor fraction of Ab produced during malarial infection is parasite specific.

Hypergammaglobulinemia, produced at least in part by polyclonal B cell activation, is a "hallmark" of acute malaria infection,[17,91] but its significance in protection and/or pathogenesis of malaria remains unresolved.

I. IMMUNOGLOBULIN ISOTYPE EXPRESSION

The class and subclass (isotype) distribution of both polyclonal and antigen-specific Abs induced by malarial infection have been studied in mice and man. Abs of the IgM, IgG_1, IgG_{2a}, IgG_{2b}, and IgG_3 isotype are produced by mice and IgM, IgG_1, IgG_2, IgG_3, and IgG_4 in man following malarial infection. Antimalarial IgA has been reported in some systems, but not in others, and in general appears to be a minor component of the overall response. IgE and IgD Ab secretions have not been implicated in malarial immunity.

TABLE 2
Summary of the Distribution of the Plaque-Forming Cell (pfc) Response by Isotype in Malaria-Infected Mice

Parasite	Mouse strain	Day of infection	% pfc Response						Ref.
			IgM	IgG$_{2a}$	IgG$_{2b}$	IgG$_3$	IgG$_1$	IgA	
Lethal	BALB/c	0	85	8		ND	5	3	93
Plasmodial yoelii		6	86	9		ND	3	2	
Nonlethal	C$_3$HeB/FeJ	0	68	20	5	4	2	ND	
P. yoelii		15	15	52	20	5	8	ND	94
Nonlethal	C$_3$HeB/FeJ	0	68	20	5	4	2	ND	
P. chabaudi		15	32	32	26	4	0.5—2	ND	94
Lethal	BALB/c	0	73	9	8	3	7	ND	
P. chabaudi		a	18	41	26	4	11	ND	95
	nu/nu	0	99	0.1	0.2	0.5	0.2	ND	
		2° infection	39	29	2	2	29	ND	

Note: ND = not determined.

a   Various numbers of parasites were inoculated (10^3 to 10^6) and pfc response measured when parasitemias reached 20%.

1. In Mice

Several studies have examined the isotype distribution of polyclonal Ig in mice infected with lethal or nonlethal *P. yoelii* and *P. chabaudi*.[91-95] Results from these studies have been recalculated by the author for comparative purposes and are summarized in Table 2. In all studies, there was an observed 30- to 100-fold increase in the total number of pfc during primary malaria infection; that is, a range of 0.2 to 5 × 10^5 total pfc per spleen were recorded for normal mice. This number rose to 1-8 × 10^6 spleen in infected mice. Thus, as many as 1/200 nucleated splenocytes may be secreting Ig during the peak of malaria infections in mice.

Although an increase in all Ig classes is observed during infection, the increase is not equally expressed among the isotypes (Table 2). The exception to this rule is lethal *P. yoelii* infection in BALB/c mice. Rosenberg[93] found an almost parallel increase in IgM, IgG$_1$, IgG$_2$, and IgA during the infection. Total IgG$_1$ pfc constituted only 3 to 9% of the total Ig pfc, regardless of the magnitude of the response which was usually 50 to 100 times above background. The host immune response to lethal *P. yoelii* was similar to that observed for other polyclonal T cell activators such as Con A.[92,93]

Like lethal *P. yoelii*, nonlethal *P. yoelii* induced large increases in secreting cells of all isotypes, but IgG$_{2a}$-secreting cells were increased out of proportion to those of the other Ig classes (Table 2). The IgG$_{2a}$-pfc response dominated the response with a 250 times increase in 15 d.[94] A similar increase in IgG$_{2b}$-pfc response was also observed. Only a small portion (~2%) of the Ig response was IgG$_1$. Thus, polyclonal isotype expression differs between lethal and nonlethal *P. yoelii* infection. One could argue that this finding was not due to the parasite per se, but to differences in the hosts (BALB/c vs. C$_3$H). This cannot be ruled out, but it seems unlikely as the Ag-specific isotype response to 17XNL *P. yoelii* is similar in the two hosts (see below), and the polyclonal isotype distribution usually parallels Ag-specific expression. Antigenic differences between the lethal and nonlethal strains of *P. yoelii* have not been found. Thus, it is rather surprising that such a small antigenic difference in the parasite results in major differences in isotype expression.

Similar but opposite results were seen in *P. chabaudi* infections. In mice infected with the lethal strain, IgG$_{2a}$ with lesser amounts of IgG$_{2b}$ predominated.[95] An approximately equal amount of IgM, IgG_{2a}, and IgG_{2b} were expressed during nonlethal infection. Surprisingly, background or below-background levels of IgG_1 were present, suggesting active suppression of the IgG_1 response.[94] Polyclonal activation in *P. chabaudi*, like with *P. yoelii*,[17,65,91] appears to be under T cell regulation as only low levels of IgG Abs were found in *P. chabaudi*-infected nu/nu mice.

During secondary *P. chabaudi* infection, isotype distribution differed greatly from that observed during primary. Approximately equal IgG_{2a} and IgG_1 responses were observed with moderate amounts of IgM and suppressed IgG_{2b} responses.[95] These polyclonal responses are similar to those observed for secondary Ag-specific responses (see below).

The Ig isotype distribution of malaria-specific Abs have been studied in mice with primary *P. yoelii*[65,96,97] and *P. chabaudi*[82] infections. In BALB/c and C_3H mice infected with *P. yoelii*, a transient anti-*P. yoelii* IgM response was observed, followed by a predominant IgG_2, a substantial IgG_3, and a low, slow-developing IgG response.[82,97] In contrast, *P. chabaudi* infections were characterized by predominant IgM responses, moderate IgG_2 and IgG_3, and little significant IgG_1 response during a primary infection.[82] The predominant isotype patterns observed in the Ag-specific response were similar to that of the polyclonal response. The exception to this is perhaps the high antiparasite IgG_3 response in the absence of polyclonal IgG_3. Studies show that at least 50% of the protein antigens present in *P. yoelii* and *P. chabaudi* are shared by both parasites. Although these two species of malarial parasites produce nonlethal infections and share many Ags in common, they do not induce similar Ab responses. This is reminiscent of polyclonal isotype differences between lethal and nonlethal *P. yoelii* and *P. chabaudi* described above. Thus, it appears that it is the "basic biology" of the parasite and not its antigenic composition that modulates isotype expression.

As would be predicted, the genetic composition of the murine host influences the course of infection and the malarial isotype expression.[97] In a recent study, 11 inbred strains of mice were infected with 17XNL *P. yoelii*, and the cause of parasitemia and kinetics of antimalarial Ab production were followed using an isotype-specific assay. Severity of infection differed among the strains. All mice developed similar, rapid IgM responses, but only 3/11 inbred strains produced significant IgG_1 levels during primary infection. All strains produced an IgG_2 response, which developed slightly more quickly in strains with the least severe courses of malaria. In general, IgG_3 Abs were the first IgG isotype to appear in serum. They were detected as early as day 8 in strains that developed mild infections and were not present until around day 20 in strains with the most severe cases of malaria. Thus, the genotype of the host influenced the time course (kinetics) of antimalarial Ab production during primary infection, but not isotype expression; that is, all mice produced comparable levels of IgM, showed generally poor to absent IgG_1 responses, predominant IgG_{2a} responses, and varying but sizable levels of IgG_3.

The classes of antimalarial Ab produced following secondary infection to *P. yoelii* and *P. chabaudi* differ from those produced during primary infection (Table 3). High levels of IgM and IgG_3 Abs are present day 15 after secondary nonlethal *P. chabaudi adami* infection, with small amounts of IgG_1 and IgG_2 reported.[82] *P. yoelii* differs from *P. chabaudi* in that IgG_2 and IgG_3 classes predominate, with increasing amounts of IgG_1 produced and low IgM levels. As shown in Table 3, the Ab response following secondary *P. yoelii* was similar among all 11 inbred strains of mice tested. It is particularly interesting that during primary infection, only 3/11 strains produced antimalarial IgG_1 Abs, but all strains produced Abs of this isotype on secondary exposure. It appears that IgG_1 responses are "down regulated" during primary infection, and that isotype expression, both polyclonal and Ag specific, are mediated by different immunologic mechanisms during primary and secondary exposure.

In immunization/vaccine studies, the isotype of antimalarial Abs induced varies when extracts of *P. chabaudi* are administered with different adjuvants.[16] BALB/c mice were

TABLE 3
Antimalarial Antibody Response Following Secondary Infection

Parasite	Mouse strain	Log₁₀ antimalarial Ab titers following secondary infection				Ref.
		IgM	IgG₁	IgG₂	IgG₃	
Plasmodial chabaudi	C₃H	3.3	1.6	2.0	3.2	82
P. yoelii	C₃H	<1.0	3.0	3.2	2.8	82
	B10.Br	1.8ᵃ	3.2	3.9	3.7	Taylor
	C₃H					
	BALB/c					
	NZB × NZW	2.3	2.6	3.9	3.9	Taylor
	DBA/2					
	B6(H—2ᵏ), B10.D2					
	AKR, B6, B10	2.0	2.0	4.3	3.5	Taylor
	B6.TL					

ᵃ Mean for indicated strains of mice.

immunized twice with a crude extract of *P. chabaudi* emulsified in either complete Freund's adjuvant (CFA), Freund's incomplete adjuvant (FIA) + pertusis, pertusis alone, or saline. Analysis of sera following immunization showed low antimalarial Ab titers of all isotypes in animals immunized with Ag alone or Ag + CFA. In animals receiving Ag + pertusis, a dominant IgG_{2a} response with some IgG_1 and IgM and no IgG_{2b} was seen. The addition of FIA to Ag + pertusis resulted in large amounts of IgG_{2a} and IgG_{2b} and suppressed IgG_1 responses. Animals in the latter two groups showed the greatest reduction in parasitemia (i.e., protection) upon challenge. Thus, expression of both correct idiotype and isotype may be important in protective immunity.

Immunologic mechanisms that led to the expression of a predominant IgG subclass in mice are unclear. Currently it appears that there are at least two T_H cell subsets in mice.[98] One T_H subset secretes IL-4 (BSF-1), which results predominantly in the production of IgG_1 and IgE, and the suppression of IgG_3, IgG_{2b},[99] IgG_{2a}, and IgM.[100] The other subset secretes INF-γ which enhances IgG_{2a} and suppresses IgG_3, IgG_{2b}, as well as IgG_1 and IgE.[100] High levels of interferon can be found in the serum of mice with *P. berghei* infection.[101] High interferon levels would therefore help account for the high IgG_{2a} and suppressed IgG_1 seen in 17XNL *P. yoelii*. Future studies on the role of these, and other B cell factors, and information on the interaction of various T_H cell subsets with malarial antigens will help our understanding of isotype expression in malarial immunity.

2. In Humans

As in mice, a large increase in polyclonal and malaria-specific Abs occurs in man following primary infection.[78,80] Both polyclonal and antimalarial Ab responses were followed during primary *P. vivax* infections in prison volunteers (Figure 6). A close parallel between the polyclonal and Ag-specific response was observed. A large increase in total Ig was seen in the volunteers, with a mean increase in IgM of 3.5 mg/ml; IgG, 4.8 mg/ml; and IgA, 1.3 mg/ml.[78] Primary Ab responses have also been measured in infants with *P. falciparum* infections.[102] In a longitudinal study, 31 Ghanaian infants were followed monthly from birth for sera conversion. Following conversion, Ab titers peaked at the 1st month but were generally low (mean geometric titer <1:80) and declined to undetectable levels within a few months.

Individuals who have been repeatedly exposed to malaria also show high levels of total circulating Ig. In a detailed longitudinal study, McGregor and Giles[103,104] followed children and cohorts living in the same village in Gambia and measured total serum Ig levels from

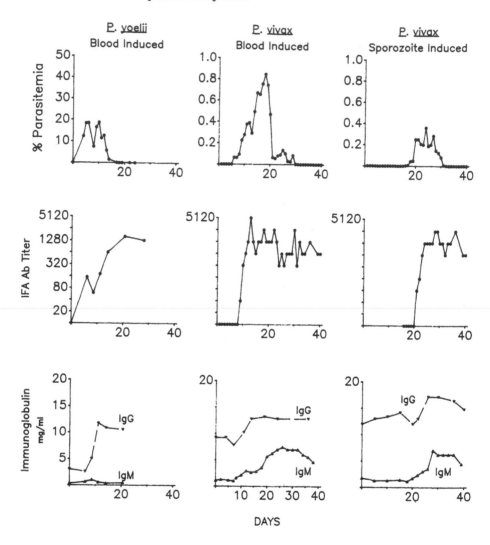

FIGURE 6.   Comparison of kinetics of parasitemia and antibody responses in mice and man. In the first panel, a group of BALB/c mice was injected with 10^6 17XNL *P. yoelii* parasitized erythrocytes. In the middle panel, a human volunteer was injected with 12.9×10^6 *P. vivax*-infected erythrocytes and in the last panel, *P. vivax*-infected mosquitoes were allowed to feed on a human volunteer. Human volunteers were treated with chloroquine days 30-32. Antimalarial Ab titers were measured by indirect immunofluorescence and Ig levels by quantitative radial diffusion. Data for this comparative figure provided by C. B. Evans. (For details, consult references 78 through 80.)

birth through age 6 years. At birth, the proportion of serum Ig (percent total protein) was higher than that normally found in European children. A portion of the children were chemotherapeutically protected from malaria, whereas the others were treated only upon infection (exposed group). There was no difference in Ig levels (i.e., percent Ig of total serum proteins) between "protected" and exposed children from the first year of life (19.4 vs. 19.7%), but a significant difference was found between the two groups at 2 years of age (20.9 vs. 26.1%).[103] At age 6, exposed children still had higher levels (25.2 vs. 31.4%) and these levels were comparable to those found in adults residing in the population.[104]

Cohen and Butcher[37] compared the rate of Ig synthesis in naturally exposed adult with adults (residing in the same area) protected by prophylaxis. Their rates of albumin synthesis were the same, but Ig synthesis was about sevenfold greater in exposed individuals, demonstrating increased Ig synthesis even in adults. Undoubtedly, the majority of this was not parasite specific.

The subclass distribution of anti-*P. falciparum* Abs has been determined by Wahlgren and colleagues[105,106] in Swedish patients with primary infection, Colombians with acute infections, and children (ages 2 to 15) residing in an area of holoendemic malaria in Liberia.[105] Anti-*P. falciparum* IgM and all four human IgG isotypes could be detected, but an interesting pattern of reactivities was seen. By IFA, Abs of all isotypes gave bright parasite reactivity, but IgG_2 Abs often gave straining restricted to the surface of schizonts. In sera with high Ab titers, Abs of all four Ig isotypes were present, sera with lower titers contained predominantly IgM, IgG_3, and IgG_1 and little IgG_2 or IgG_4. In overall immune response, there was a relationship between the presence of Abs of each isotype, and the gene (Igh-C) sequence of DNA in man ($5'$ . . . μ, $\gamma3$, $\gamma1$, $\gamma2$, $\gamma4$, . . . $3'$); that is, if IgG_4 Abs were detected, then Abs of the other three IgG classes (coded for by upstream genes) were also present. Sera with high IgG_3 and IgG_1 levels were found which lacked (or had low levels) of IgG_2 and IgG_4. Relationship of Ig expression and heavy-chain DNA sequence is similar to that found by Mongini et al.[107] for the T-independent antigen TNP-ficoll in mice. The relations between Ab isotype expression and gene sequence are probably not a generalized phenomena of malaria, as Ab isotype expression in rodent malaria does not show this relationship.[82,97]

3. Possible Significance of Ig-Isotype Expression

It has been shown in several infectious diseases that some Ig isotypes are more efficacious than others in mediating protection. Similar studies have not been reported previously in human or mouse malarias. We collected hyperimmune sera from BALB/c mice exposed several times to *P. yoelii*, and separated the sera into various fractions (IgM, IgG_1, IgG_{2a}, IgG_{2b}, and IgG_3) by Protein A-chromatography. Each purified Ig fraction was concentrated until the titer of the fraction was equivalent to the titer of the corresponding isotype in hyperimmune sera. The protective value of each Ig isotype was assessed in passive antibody transfer studies. In three repeat experiments, transfer of IgM, IgG_1, IgG_{2b}, and IgG_3 did not alter the course of lethal *P. yoelii* infections, whereas IgG_{2a} was as effective as hyperimmune sera. These results suggest that antimalarial IgG_{2a} Abs are of an appropriate concentration and isotype to mediate protection, whereas the other isotypes are not. Since IgG_{2a} Abs are in the highest concentration in hyperimmune serum, it is possible that the other classes could mediate protection if they were present at higher concentrations. To test for this, hyperimmune anti-*P. yoelii* sera from a large number of mice were fractionated by Protein A-chromatography, and each purified Ig isotype concentrated until it reached the level of IgG_{2a} found in hyperimmune sera. Upon passive transfer, IgM and IgG_1 were ineffective, IgG_3 was partially effective, and IgG_{2a} was again as effective as hyperimmune sera in mediating immunity. Thus, IgM and IgG_1 concentrated to ~500 times that normally found in hyperimmune sera were still not able to mediate protection. The IgG_3 fraction was contaminated with low levels of IgG_{2a}, so further studies will be required to assess the protective potential of IgG_3 Abs. It is of interest that mAbs that passively transfer partial immunity to *P. yoelii* are of the IgG_{2a} and IgG_3 isotypes. These results suggest that Ig-isotype expression may be important in malarial immunity. The importance of Ig-isotype expression in the pathogenesis of malaria is discussed in Section L.

J. ANTIBODY AFFINITY/AVIDITY

Generally, determination of Ab binding affinities requires the availability of purified hapten or antigen. Since, until recently, it has not been possible to obtain large amounts of purified malarial antigens, direct determination of antimalarial Ab affinity or avidity generated during primary or secondary infection has not been feasible. Although no direct measurements have been made, several pieces of evidence suggest that Abs of low affinity or avidity may be produced.

Anders[108] speculated that the presence of cross-reacting epitopes in malarial parasite could prevent selection of high-affinity clones. Many of the plasmodial antigens that have been sequenced contain tandem repeats, e.g., the CSP,[109] the S-Ag,[110] and RESA (Pf155).[111] It is hypothesized that the presence of tandem repeats could result in a large number of immunologically related cross-reactive epitopes. Common epitopes could be found on protective Ags as well as reasonably unimportant or self-Ags. Usually, during an immune response as Ag becomes limiting, B cells with high-affinity receptors (some of which will undoubtedly be generated by somatic mutation) will be stimulated, whereas those with lower affinity receptors will not. However, if there are a large number of cross-reacting Ags, they could continue to stimulate low-affinity clones. Thus, the overall serum avidity for malarial Ags would remain low. Individual Ags with multiple repeating sequences, however, could have high avidity for Abs produced by low affinity clones.

Hamburger and Kreier[112] postulated that antimalarial Abs, including those capable of mediating protection against merozoites, were of low affinity. They incubated freed *P. berghei* parasites in fixed amounts of hyperimmune rat anti-*P. berghei* sera in various volumes of diluent and then measured infectivity *in vivo*. They theorized that if high-affinity Abs were present they would rapidly bind to the parasite and remain attached. If low-affinity Abs were present and the amount of diluent was increased, more Abs would "back off" the parasite, and Ab would be in high concentration in the diluent of equilibrium. Their results support the presence of low-avidity Abs. They also reported being able to wash protective Abs from the surface of freed parasites.[113] They speculate that the low binding efficiency of protective Abs to malarial parasites may contribute to parasite evasiveness.[113]

Although tandem repeat sequences may contribute to generation of low-affinity Abs, the story may be more complicated. In 1973, Steward and Voller[114] immunized inbred Simpson mice with 1 mg of the antigen human serum transferrin (HST) when the mice had a ~2% *P. yoelii* parasitemia, and again, a week later, at ~8.9% parasitemia, 25 uninfected control mice received the same immunization. Sera were collected and Ab titers and relative affinity contents (K_R) determined 2 weeks after immunization. Surprisingly, 79% of infected mice produced anti-HST Abs, compared to only 40% of controls. Ab titers (expressed as $\mu\mu$ moles binding sites per milliliter) were significantly lower in infected mice (0.39×10^6 vs. 1.9×10^6). Since HST is not known to have repeat sequences or to cross-react with *P. yoelii* Ags, mechanisms other than high levels of cross-reactive epitopes may contribute to low-affinity Ab production, if indeed low-affinity Ab production occurs during primary malaria.

K. PERSISTANCE AND CATABOLISM OF ANTIMALARIAL ABS

Serum Ab titers decrease slowly in mice once erythrocyte-stage parasites are eliminated. In *P. yoelii*-infected CBA mice, parasites are eliminated from the peripheral blood by week 3 after blood-induced infections, but parasites can be detected in kidneys up to week 4.[64] By week 6, antimalarial IgM and IgG$_1$ titers begin to decrease, but IgG$_2$ levels remained unchanged.[76] In a separate study, we followed the kinetics of antimalarial Ab disappearance in BALB/c mice following recovery from 17XNL *P. yoelii*. A summary of Ab titers as determined by an isotype-specific assay is shown in Table 4. As can be seen, Ab titers of all classes decreased with time, IgG$_3$ titers dropping the most rapidly. Anti-*P. yoelii* Abs were still present 9 months postinfection. To determine if Abs to some Ags disappeared faster than others, sera collected 1, 3, 6, and 9 months after primary *P. yoelii* infection were used in immune precipitation studies. Results in Figure 7 show that Ag specificity of Abs present at 9 months did not differ significantly from that at month 1. Thus, it appears that decline in Abs to all detectable malaria Ag occurs at the same rate.

It is surprising that Abs persist for up to 1 year following recovery from *P. yoelii*. The half-life of murine IgG Abs ranges from 3 to 5 d; thus, once the infection is eliminated, one would expect a rapid decline in malarial Ab titers. At 6 to 9 months post infection,

TABLE 4
Antibody Titers in Plasma Following Recovery from Primary
***P. yoelii* Infection**

Postinfection (months)	Total IFA Ab titer	Log$_{10}$ RIA titers			
		IgM	IgG$_1$	IgG$_2$	IgG$_3$
1	1:2048	4.1	4.1	5.1	4.2
3	1:256	ND	ND	ND	ND
6	1:256	<1.0	3.1	3.2	1.2
9	1:128	<1.0	2.2	3.1	1.0

Note: ND = not determined.

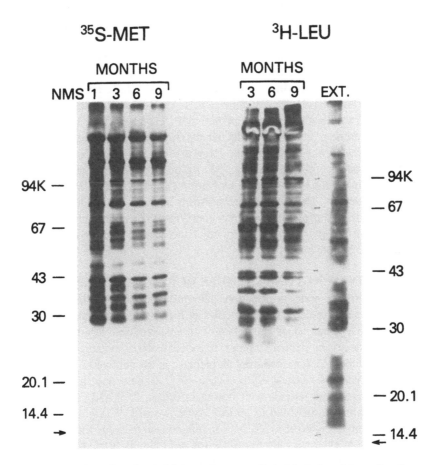

FIGURE 7. Detection of malarial Ags using sera collected 1, 3, 6, and 9 months after primary *P. yoelii* infection. *P. yoelii* parasites were metabolically labeled with ^{35}S-MET and ^{3}H-LEU, solubilized in NP-40, and used in immunoprecipitation studies. Equivalent amounts of labeled Ag and Abs were used in each assay.

spleens appear dark, suggesting the continued presence of malarial pigment. It is also possible that plasmodial Ag is retained by dendritic cells. The mechanism for continued Ab production to a large number of malarial Ags requires further study.

In humans, Ab titers appear to rapidly decrease following elimination of the parasite. In prison volunteer *P. vivax* studies, Ab titer (as determined by indirect IFA) dropped from a peak of 1:5120 to 1:20 to 1:40 by day 150.[80] However, the level of infection was controlled chemotherapeutically, so maximal immune responses may not have been obtained in these individuals. Similar results were found in *P. falciparum* volunteer studies.[81] Wyler and Oppenheim[115] collected blood from individuals infected 1 month to 28 years with *P. falciparum*. Some of those individuals had only experienced a single episode of malaria and had been cured chemotherapeutically. Sera collected at 1 to 7 months had low IFA Ab titers (mean 1:66), whereas those collected >1 year were serologically negative.[115] Ab titers also drop rapidly in immune individuals who move to nonmalarious areas.

Cohen et al.[3] measured the turnover rate (catabolism) of Ig in Europeans, West Africans living in Great Britain, and West Africans exposed to malaria. Subjects were injected with ^{131}I-albumin or [125]-γ globulin and rate of elimination was followed for ~1 month. There was no difference between the groups in the catabolism of albumin, but the rate of globulin depletion was significantly increased in malaria-exposed West Africans (169 mg/kgm/d) compared to West Africans in England (50 mg/mgm/d) and Britons (23 mg/kgm/d).[3] Thus, the turnover rate of Ig in exposed/infected individuals for malaria-specific and -nonspecific Ig is enhanced.

The above results suggest that malarial infection may enhance catabolism of Abs to other pathogens or Abs acquired by vaccination. Goumard and colleagues[90] immunized mice with the T-dependent Ags, bovine serum albumin (BSA), HST, tetanus toxoid (TT), and the T-independent Ag polyvinylpyrrolidone (PVP). Then 15 d after the last immunization, a portion of mice was challenged with nonlethal *P. chabaudi,* and 15 d later (following clearance of the parasites), sera were collected and Ab binding capacity (ABC) determined. A significant decrease in ABC was found for all Ags in *P. chabaudi*-recovered mice, especially BSA (2600 ng/ml in control vs. 500 ng/ml infected) One might have speculated the titer might increase due to polyclonal activation (which is extensive in this system), but a decrease was observed. The precise mechanism is unclear, but the authors speculate that enhanced Ig catabolism during infection may contribute to this phenomenon. Thus, malarial infection may have a negative effect on maintenance of Ab-mediated immunity achieved through childhood immunization.

L. PRODUCTION OF AUTOANTIBODIES DURING MALARIAL INFECTION

During the course of malarial infection, autoantibodies are produced in virtually every plasmodial host system studied. Autoantibodies of highest interest have been those directed against normal erythrocytes, since the extent of anemia produced is greater than that attributable to erythrocyte loss by parasite rupture. In rodent malaria, evidence for anti-erythrocyte Ab production includes (1) demonstrating an increase in the number of splenic pfc secreting Abs to bromelin-treated, syngeneic erythrocytes,[77,91,93] (2) detecting Ig bound to the surface of normal erythrocytes and reticulocytes during infection,[116-118] and (3) identifying Abs in sera of recovered animals that bind to normal erythrocytes.[116,117] Some of the Abs are cold reactive type and can be eluted at 37°C.[116,118] In infected animals, the kinetics of Ab secretion to "modified" host erythrocytes parallels the polyclonal Ig response.[76,77,91,116] Abs to normal erythrocytes can be detected as early as day 2 in some systems.[117] In general, Abs of the IgG class are involved with IgM Abs detected in some animals. Parasitized and nonparasitized reticulocytes have been shown to have larger amounts of membrane-bound Ig than mature, nonparasitized erythrocytes.[116] Several investigators have reported that there is no enhancement of autoerythrocyte production during secondary infection. Cold agglutinins can be detected, but their possible role in anemia remains speculative.

A large proportion of humans infected with *P. falciparum* develop autoantibodies. The thorough investigations of C. Facer[119-121] show that ~50% of *P. falciparum*-infected Gambian children had host-serum components associated with the surface of normal erythrocytes.[119] The most frequent form of erythrocyte sensitization was C3d, but IgGC3d and IgGC3bC3d were also present. In a study of 20 children with erythrocyte-bound IgG, 13 had IgG$_2$ or

IgG_4, either alone or in combination.[121] These children were never found to be anemic. However, a positive correlation was found between anemia and sensitization with IgG_1. In addition, there was a greater chance of finding a child anemic if IgG_1 were present and IgG_4 were absent. IgG_3 Abs and IgG_1 Abs in humans are good and moderately effective, respectively, in activating the classical pathway of complement (IgG_2 is only slightly effective and IgG_4 does not bind C1q), and human monocytes have receptors only for IgG_1 and IgG_3).[121] Thus, IgG_1 and IgG_3 Abs would be the most effective in mediating anemia through hemolysis and erythrophagocytosis.

In addition to antierythrocyte Abs, Abs to other cell types have been reported in mice and man.[122] Antilymphocytotoxic Abs were found in the sera of 95% of *P. falciparum* and 98% of *P. vivax* patients when assayed at 15°C.[123] Activity at 37°C was significantly less than that at 15°C. These "cryo" antilymphocytotoxic Abs were IgM, had titers of 1:2 to 1:16, and were directed primarily against B cells and to a lesser extent against T cells and macrophages.[124] The role of these Abs in immunoregulation remains unresolved.

The mechanisms of autoAb activation is not clear. It has been speculated to result from polyclonal B cell activation, parasite-modification of normal erythrocytes possible by the release of enzymes, the attachment of soluble parasite antigens to erythrocyte surfaces, Ab induction by cross-reactive plasmodial antigens, malaria-associated suppression of T_S cells activity,[125] and the adherence of immune complexes to the erythrocyte surface.[116-119] Acid elution of erythrocyte surface-bound Ig has been shown to release both Abs that detect malarial antigens and malarial Ags themselves, thus supporting the hypothesis of immune complex involvement and particulation of malarial Ags.[120] Ronai and Sulitzeanu[117] reported that adsorption of antierythrocyte activity also removed antilymphocyte reactivity in a rodent model supporting the notion of cross-reactive Abs. In many of the experiments malarial Ags and immune complexes are clearly present; thus, antierythrocyte Abs (autoAbs) may not always be involved.

Some investigators have hypothesized that "autoantierythrocyte antibodies" may be involved in anemia, whereas others have speculated that they may constitute a defense mechanism of the host against the parasite.[76,123] Abs to modified erythrocyte Ags, which are present on infected cells, could aid in the clearance and removal of infected cells through a variety of methods including erythrophagocytosis.[126] Much is to be learned about the mechanism of induction and role of autoantibodies in anemia, immune regulation, and protective immunity.

III. BRIEF SUMMARY OF THE ROLE OF ANTIBODIES IN SPOROZOITE IMMUNITY

A. EVIDENCE THAT AbS ARE IMPORTANT IN SPOROZOITE IMMUNITY

Animals become protected from challenge with viable sporozoites if they are first immunized with attenuated sporozoites[127-129] or with viable sporozoites followed by drug treatment.[130] Following immunization, high antisporozoite Abs titers develop.[127] A correlation between protection and prechallenge Ab titers has been reported in some studies.[129] Mice immunized with frozen sporozoites, however, fail to develop antisporozoite Abs and do not develop protective immunity.[129]

Sera with high antisporozoite Ab titers can modulate infection. If sporozoites are incubated *in vitro* in immune sera and then injected into naive recipients, the infectivity of the parasite is significantly reduced or abolished.[131] Infectivity is not significantly reduced by normal sera. Thus, immune sera contains sporozoite neutralization activity (SNA). Abs in this sera bind to the surface of sporozoites and produce a long trail of Ag-Ag extending from the parasite. This is called the circumsporozoite (CS) reaction.

If large amounts of immune sera are passively transferred to animals at the time they are injected with viable sporozoites, an increase in the rate of sporozoite clearance from the circulation results and a reduction in the number of exoerythrocytic stages in the liver is observed.[132] However, unlike in immunization studies where complete protection can be achieved, complete protection is rarely found with the passive transfer of Abs.[133] Spitalny et al.[134] injected mice intravenously once with 0.9 ml or three times with 0.5 ml of sera with high sporozoite neutralizing activity and CSP reactivity, and then challenged the recipients with only 5000 viable sporozoites. A mean of 4.8 d later, 17 of 18 Ab-treated mice developed malaria. This did not differ significantly from 17/17 control mice treated with normal sera (mean of 4.1 d).

The target Ag on the sporozoites surface that mediates sporozoite neutralization and the CS reaction (called the CS protein—CSP) has been identified using mAbs,[135-138] and the gene coding for it,[109] cloned and sequenced. The CSP is approximately 40,000 to 50,000 Da (weight varying with the species of parasite being studied),[135] and has a series of tandem repeatsthroughout the internal portion of the molecule.[109] MAbs, produced against the CSP protein, mediate the CS reaction *in vitro*.[136,137] The passive transfer of 100 μg, and in some experiments 10 μg, of mAb to the CSP protected mice from challenge with 1000 viable sporozoites.[136] In this case, the efficiency of mAbs was even greater than immune sera. In additional studies, sporozoites were incubated *in vitro* with various amounts of mAb or Fab fragments and then 10^4 sporozoites were injected into mice. As little as 48 μg/ml of Ag or Fab was sufficient to completely neutralize sporozoite infectivity.[136] It is clear from these studies that Ab to a single epitope can prevent an animal from developing malaria.

In a series of *in vitro* studies, mAb to the CSP has been shown to prevent sporozoites from attaching and entering target cells (in this case W138 human lung cells).[139] Hollingdale et al.[139] incubated *P. berghei* sporozoites with varying amounts of purified mAb or Fab fragments. When 20,000 to 40,000 sporozoites were incubated with 15 to 100 μg/ml of mAb, and then added to monolayers of W138 cells, intracellular trophozoites were not seen 24 h later. Control W138 cells, treated with an equivalent amount of irrelevant Ab, contained large numbers of developing trophozoites. If lesser amounts of Ab were used, a significant decrease in the number of intracellular parasites was still seen. However, those sporozoites that were developing appeared morphologically normal at 48 h. Thus, it was concluded that the Abs probably functioned to prevent the attachment and entry of the sporozoite into host hepatic cells, but were ineffective once the parasite was inside.

In conclusion, high Ab titers to the CSP are produced as animals become functionally immune during immunization with irradiated sporozoites, they bind to the surface of the parasite, and block the penetration of sporozoites into cells *in vitro*. However, the correlation between Abs titers and protective immunity to sporozoites is certainly not absolute. The level of immunity observed by passive Ab transfer studies is never as good as that found in immunized mice, and in immunization studies protective immunity preceeds the appearance of detectable serum Abs.[128]

Evidence for an immune mechanism, other than Abs, was elegantly shown by Chen et al.[140] using μ-suppressed mice. B cell deficient and intact BALB/c mice were immunized with irradiated sporozoites. High titers of anti-CSP Abs were present in intact animals and completely absent in μ-suppressed animals. When challenged with 10^4 viable sporozoites, 7/11 (64%) of B cell deficient mice were protected, compared to 100% of immunized controls and 0% of unimmunized animals. This demonstrates that, in the absence of Abs, immunity to sporozoites can be induced.

Further evidence that immune mechanisms, other than Abs, are important comes from epidemiologic studies showing minimal correlation between anti-CSP Ab levels in humans and protection.[141] Ab levels seem to correlate with exposure rates.[142,143] Children receive Abs from their mothers.[144] After neonatal immunity wanes, children living in hyperendemic areas produce Abs to the CSP protein by 3 years of age, with adult levels being reached by

10 to 19 years of age (Figure 4). However, these children are still susceptible to infection. Recently, Hoffman and colleagues[141] compared the disappearance of anti-CSP Abs and susceptibility to *P. falciparum* malaria in 83 adults living in Kenya. Subjects were treated with Fansidar to verify the absence of parasites, and then Ab titers to the CSP protein and the development of malaria due to natural environmental exposure were followed. Within 98 d, 72.3% of the individuals develop malaria. There was no significant difference in CSP Ab titers between those that got malaria and those that did not.[141] Webster et al.[145] studying Thai individuals with acute *P. falciparum* malaria, found that Ab titers to the CSP were boosted during infection, but that titers declined rapidly following chemotherapy. They found that anti-CSP Ab had a circulating half-life of about 27 d (which is normal for IgG), and that the levels of antisporozoite Abs did not increase during the erythrocytic stage of infection. They suggest that immunosuppression is associated with malaria and that this may prevent the acquisition of higher Ab titers.

Thus, it appears that Abs are a reasonable indicator of exposure, that they are effective in *in vitro* studies, but their value in controlling natural infections is unclear. It is likely that Abs help decrease the number of sporozoites that enter hepatocytes, but once within hepatocytes other immune mechanisms are relevant in controlling exoerythrocytic stage development.[146] It is also unclear if Ags, other than the CSP, are relevant in antisporozoite immunity.[147]

B. MECHANISMS OF AB PRODUCTION

Information on mechanisms of Ab production against the CSP protein and other sporozoite Ag is limited. With the availability of synthetic peptides of the CSP Ag, current methodology in cell separation, and Ag-specific plaque-forming cell assays, this information should be forthcoming.

The spleen is the major site of Ab production. When mice are splenectomized and then immunized with irradiated sporozoites, only 15 to 20% of the mice developed positive CSP reactions with little or no sporozoite neutralizing activity.[134]

Following exposure to viable *P. berghei* sporozoites, anti-CSP Abs can be found in the sera of intact rats by 6 to 7 d, titers reach a peak about day 14, but decline rapidly thereafter.[128] Animals immunized with irradiated sporozoites, develop peak Ab titers at approximately day 14, but titers remain elevated for several months. As mentioned above, anti-CSP and antisporozoite Abs decline rapidly in man.

Ab production to sporozoites is under T cell regulation.[148] When thymectomized, lethally irradiated, bone marrow-reconstituted mice are immunized repeatedly with attenuated sporozoites, 50% of the mice produce weak CSP titers (1:2 to 1:10), the other 50% are totally nonresponders. The authors feel the marginal Ab titers in some mice may be due to residual T cells in thymectomized hosts.[148] The CSP contains repeat epitopes, but there is no evidence that it functions as a T-independent Ag. Like Ab production to the erythrocytic stage, Ag processing by macrophages (or possibly Kupffer cells) and T cell activation appear to be required for Ab production.

Induction of Abs to the CSP Ag is under MHC restriction in mice[149] and presumed to be so in humans.[150] Good et al.[149] immunized mice with a 24 amino acid segment of the tandem repeat of the CSP. Only mice with the H-2^b haplotype I-A^b produced Abs. These results indicate that the 24-mer contained a T cell epitope capable of being seen by mice expressing an I-A^b molecule, as well as a B cell epitope. To determine the length of the T cell epitope, primed lymphocytes were stimulated *in vitro* with various lengths of the CSP repeat sequence. Optimal response was with a 16-mer, moderate response with 11-mer, but no response was found with a 7-mer. This work has been confirmed and expanded.[150]

If the ability to recognize the CSP Ag of *P. falciparum* is MHC restricted in man, one may observe that within a village, some families would produce Abs while other families would not. Del Giudide et al.[151] reported considerably different Ab titers to the CSP repeat region in children living in different households, but exposed to the same epidemiological conditions and with similar Ab titers to the erythrocytic stages. Further studies are needed to determine the restricting element in man.

Following repeated exposure, humans produce Abs of all four IgG isotypes to *P. falciparum* CSP with IgG_1 and IgG_3 predominating.[141] It is not known if polyclonal activation accompanies the Ag-specific response to sporozoites. Information on the isotype distribution in mice to the rodent plasmodial CSP is not available.

IV. IMPORTANCE OF AbS IN IMMUNITY TO GAMETOCYTES AND GAMETES

During the intraerythrocytic developmental cycle, some parasites differentiate into macrogametocytes and microgametocytes. Like the other intraerythrocytic stages, they are protected from "direct Ab attack" by the host erythrocyte membrane. In some species, e.g., *P. falciparum,* gametocytes continue to circulate in the peripheral blood after the other erythrocytic stages have cleared. It has been reported by various investigators that Abs do not appear to be effective against circulating gametocytes.[3]

When mosquitoes feed on infected hosts, they ingest gametocytes. Within the midgut of the mosquito, the parasites are released from erythrocytes and develop sequentially into gametes (male and female), zygotes, and ookinetes. In 1976, Gwadz[152] and Carter and Chen[153] showed that fertilization and further parasite development were prevented within mosquitoes that had fed on chickens immunized with irradiated gametes or gametocytes of *P. gallinaceum.* In one set of experiments, chickens were immunized with 2.7×10^5 to 2.2×10^7 male gametes, challenged with *P. gallinaceum,* and then mosquitoes were allowed to feed daily during patent parasitemias.[153] A total of 0 to 1 oocysts was found in the midgut of mosquitoes fed on vaccinated birds compared to 2267 oocysts in those fed on control, unvaccinated birds. The process of preventing gamete development is known as *transmission blocking immunity* and had been demonstrated for *P. yoelii,*[154] *P. falciparum,*[155] and *P. vivax.*[156] It is clear that Abs present in the sera of the host mediate transmission blocking immunity within mosquitoes.[152,155,157,158]

Rener et al.[155] showed that a pool of two mAbs, directed against the surface of *P. falciparum* gametes, caused rope-like agglutination and this prevented microgametes from fertilizing macrogametes. These mAbs bound to the surface of both male and female gametes and identified three surface iodinated proteins of M_r 255, 59, and 53 kDa under reducing conditions (235, 49, and 46 kDa under nonreducing). Subsequent studies by Vermeulen and colleagues[158] confirm that Abs to the 45/48 kDa Ags act by preventing or reducing fertilization of female gametes. The involvement of the high molecular weight protein (230 to 255 kDa) remains less clear.

In addition to the gametes, zygotes and ookinetes are accessible to Ab interaction within the mosquito. A 25-kDa Ag has been identified on the surface of maturing macrogametes, zygotes, and ookinetes of *P. falciparum* that is glycosylated and acylated.[158,159] Abs against these two Ags severely suppress the development of the parasite within the mosquito. Anti-25-kDa Abs are active against ookinetes,[158] and may damage their surface coats, thus making them accessible to the digestive activity of the secretions of the mosquito stomach.[160] This process may be complement dependent. It has also been suggested that the M_r 25-kDa protein acts as a ligand for the receptor on the mosquito midgut cell wall and is therefore important for the penetration of the ookinete.[158]

Abs to gametes, and probably ookinetes, mediate part of their effect by complement. Rener et al.[155] reported that in the presence of complement IgG_{2a} mAb was ten times more effective than in the absence of complement. Some IgG_{2a} anti-*P. gallinaceum* mAbs, which otherwise mediated minimal or no suppressive effect, completely abolished infectivity of the parasite if complement were present.[157] Thus, transmission-blocking Abs mediated their effects by agglutination, complement-mediated lysis, and possibly by preventing penetration.

In nature, humans produce transmission-blocking Abs during primary *P. vivax* infection.[156] If sera with transmission-blocking activity is heat inactivated to destroy complement, ~50% of the activity is lost.[156] Thus, in man both complement-dependent and -independent Abs mediate the effect. Since transmission-blocking Abs are rapidly produced during acute infection, many questions arise: How does naturally induced transmission-blocking immunity influence transmission in nature? How does the presence of transmission-blocking Abs compare with the appearance of gametocytes in the blood? What effect would vaccination provide for blocking malarial transmission? Answers to these and other questions will be of value in our understanding of this most interesting approach to the control of malaria.

V. ACKNOWLEDGMENTS

I would like to acknowledge the contribution of Charles B. Evans to this chapter. Over the last 30 years he has amassed lab books of data on the Ab response to malaria in human volunteers, various primates, and experimentally infected mice. Data collected 20 years ago, as well as 2 months ago, were taken from his lab books for inclusion in this chapter. I would also like to acknowledge the tireless efforts of T. Roberts in helping research, type, and prepare graphics for this chapter.

REFERENCES

1. **Coggeshall, L. T. and Kumm, H. W.**, Demonstration of passive immunity in experimental working malaria, *J. Exp. Med.*, 66, 177, 1937.
2. **Coggeshall, L. T. and Kumm, H. W.**, Effect of repeated superinfection upon the potency of immune serum of monkeys harboring chronic infections of *Plasmodium knowlesi*, *J. Exp. Med.*, 68, 17, 1938.
3. **Cohen, S., McGregor, I. A., and Carrington, S.**, Gammaglobulin and acquired immunity to human malaria, *Nature*, 192, 733, 1961.
4. **Diggs, C. L. and Oster, A. G.**, Humoral immunity in rodent malaria. II. Inhibition of parasitemia by serum antibody, *J. Immunol.*, 102(2), 298, 1969.
5. **Zuckerman, A. and Golenzer, Y.**, The passive transfer of protection against *Plasmodium berghei* in rats, *J. Parasitol.*, 56, 379, 1970.
6. **Phillips, R. S. and Jones, V. E.**, Immunity to *Plasmodium berghei* in rats: maximum levels of protective antibody activity are associated with eradication of the infection, *Parasitology*, 64, 117, 1972.
7. **Brown, I. N. and Phillips, R. S.**, Immunity to *Plasmodium berghei* in rats: passive serum transfer and role of the spleen, *Infect. Immun.*, 10(6), 1213, 1974.
8. **Golenser, J., Spira, D. T., and Zuckerman, A.**, Neutralizing antibody in rodent malaria, *Trans. R. Soc. Trop. Med. Hyg.*, 69(2), 251, 1975.
9. **Lourie, S. H. and Dunn, M. A.**, The effect of protective sera on the course of *Plasmodium berghei* in immunosuppressed rats, *Proc. Helminthol. Soc. Wash.*, 39, 470, 1972.
10. **Stechschulte, D. J., Briggs, N. T., and Wellde, B. T.**, Characterization of protective antibodies produced in *Plasmodium berghei* infected rats, *Mil. Med.*, Special Issue, 1140, 1969.
11. **Wells, R. A. and Diggs, C. L.**, Protective activity in sera from mice immunized against *Plasmodium berghei*, *J. Parasitol.*, 62(4), 638, 1976.
12. **Freeman, R. R. and Parish, C. R.**, *Plasmodium yoelii*: antibody and the maintenance of immunity in BALB/c mice, *Exp. Parasitol.*, 52, 18, 1981.
13. **Murphy, J. R. and Lefford, M. J.**, Host defenses in murine malaria: evaluation of the mechanisms of immunity to *Plasmodium yoelii* infection, *Infect. Immun.*, 23(2), 384, 1979.
14. **Jayawardena, A. N., Targett, G. A., Leuchars, E., and Davies, A. J. S.**, The immunological response of CBA mice to *P. yoelii*. II. The passive transfer of immunity with serum and cells, *Immunology*, 34, 157, 1978.

15. **McDonald, V. and Sherman, I. W.**, *Plasmodium chabaudi:* humoral and cell-mediated responses of immunized mice, *Exp. Parasitol.,* 49, 442, 1980.

16. **Miller, C.**, Ph.D., dissertation, U.M.I. Dissertation Service, University Microfilms International, Ann Arbor, MI, 1986.

17. **Weibaum, F. I., Evans, C. B., and Tigelaar, R. E.**, Immunity to *Plasmodium berghei yoelii* in mice. I. The course of infection in T cell and B cell deficient mice, *J. Immunol.,* 117(5), 1999, 1976.

18. **Roberts, D. W. and Weidanz, W. P.**, T-cell immunity to malaria in the B-cell deficient mouse, *Am. J. Trop. Med. Hyg.,* 28(1), 1, 1979.

19. **Grun, J. L. and Weidanz, W. P.**, Immunity to *Plasmodium chabaudi adami* in the B-cell-deficient mouse, *Nature,* 290, 143, 1981.

20. **Roberts, J. A. and Tracey-Patte, P.**, Adoptive transfer of immunity to *Plasmodium berghei, J. Protozool.,* 16(4), 728, 1969.

21. **Phillips, R. S.**, *Plasmodium berghei:* passive transfer of immunity by antisera and cells, *Exp. Parasitol.,* 27, 479, 1970.

22. **Kasper, L. H. and Alger, N. E.**, Adoptive transfer of immunity to *Plasmodium berghei* by spleen and lymph node cells from young and old mice, *J. Protozool.,* 20(3), 445, 1973.

23. **Gravely, S. M. and Kreier, J. P.**, Adoptive transfer of immunity to *Plasmodium berghei* with immune T and B lymphocytes, *Infect. Immun.,* 14(1), 184, 1976.

24. **Jayawardena, A. N.**, Immune responses in malaria, in *Parasitic Diseases, The Immunology,* Vol. 1, Mansfield, J. M., Ed., Marcel Dekker, New York, 1981, 85.

25. **McDonald, V. and Phillips, R. S.**, *Plasmodium chabaudi* in mice: adoptive transfer of immunity with enriched populations of spleen T and B lymphocytes, *Immunology,* 34, 821, 1978.

26. **McDonald, V. and Phillips, R. S.**, *Plasmodium chabaudi:* adoptive transfer of immunity with different spleen cell populations and development of protective activity in the serum of lethally irradiated recipient mice, *Exp. Parasitol.,* 49, 26, 1980.

27. **McGregor, I. A.**, Demographic effects of malaria with special reference to the stable malaria of Africa, *West Afr. Med. J.,* 9, 260, 1960.

28. **McGregor, I. A.**, Immunology of malarial infection and its possible consequences, *Br. Med. Bull.,* 28, 22, 1972.

29. **Brown, G. V., Anders, R. F., Mitchell, G. F., and Heywood, P. F.**, Target antigens of purified human immunoglobulins which inhibit growth of *Plasmodium falciparum in vitro, Nature,* 297, 591, 1982.

30. **Cohen, S. and Butcher, G. A.**, Properties of protective malarial antibody, *Immunology,* 19, 369, 1970.

31. **Mitchell, G. H., Brutcher, G. A., Voller, A., and Cohen, S.**, The effect of human IgG on the *in vitro* development of *Plasmodium falciparum, Parasitology,* 72, 149, 1976.

32. **Cohen, S., Butcher, G. A., and Crandall, R. B.**, Action of malarial antibody *in vitro, Nature,* 223, 368, 1969.

33. **Chizzolini, C. A., Dupont, A., Akue, J. P., Kaufmann, M. H., Verdini, A. S., Pessi, A., and del Giudice, G.**, Natural antibodies against three distinct and defined antigens of *Plasmodium falciparum* in residents of a mesoendemic area in Gabon, *Am. J. Trop. Med. Hyg.,* 1989.

34. **Cohen, S.**, Immunity to malaria, *Proc. R. Soc. London,* B202, 323, 1979.

35. **Taylor, D. W. and Siddiqui, W. A.**, Recent advances in malarial immunity, *Am. Rev. Med.,* 33, 69, 1982.

36. **Diggs, C. L. and Osler, A. G.**, Humoral immunity in rodent malaria. III. studies on the site of antibody action, *J. Immunol.,* 114(4), 1243, 1975.

37. **Cohen, S. and Butcher, G. A.**, Serum antibody in acquired malarial immunity, *Trans. R. Soc. Trop. Med. Hyg.,* 65(2), 125, 1971.

38. **Quinn, T. C. and Wyler, D. J.**, Mechanisms of action of hyperimmune serum in mediating protective immunity to rodent malaria *(Plasmodium berghei), J. Immunol.,* 123(5), 2245, 2979.

39. **Miller, L. H., Aikawa, M., and Dvorak, J. A.**, Malaria *(Plasmodium knowlesi)* merozoites: immunity and the surface coat, *J. Immunol.,* 114(4), 1237, 1975.

40. **Brown, J., Whittle, H. C., Berzins, K., Howard, R. J., Marsh, K., and Sjoberg, K.**, Inhibition of *Plasmodium falciparum* growth by IgG antibody produced by human lymphocytes transformed with Epstein-Barr virus, *Clin. Exp. Immunol.,* 63, 135, 1986.

41. **Hudson, D. E., Miller, L. H., Richards, R. L., David, P. H., Alving, C. R., and Gitler, C.**, The malaria merozoite surface: a 140,000 m.w. protein antigenically unrelated to other surface components on *Plasmodium knowlesi* merozoites, *J. Immunol.,* 130, 2886, 1983.

42. **Freeman, R. R., Trejdosiewicz, A. J., and Cross, G. A. M.**, Protective monoclonal antibodies recognising stage-specific merozoite antigens on a rodent malaria parasite, *Nature,* 284, 366, 1980.

43. **Majarian, W. R., Daily, T. M., Weidanz, W. P., and Long, C. A.**, Passive immunization against murine malaria with an IgG$_3$ monoclonal antibody, *J. Immunol.,* 132, 3131, 1984.

44. **Boyle, D. B. F., Newbold, C. I., Smith, C. C., and Brown, K. N.**, Monoclonal antibodies that protect *in vivo* against *Plasmodium chabaudi* recognize a 250,000-dalton parasite polypeptide, *Infect. Immun.,* 38, 94, 1982.

45. **Fandeur, T., Dubois, P., Gysin, J., Dedet, J. P., and da Silva, L. P.,** *In vitro* and *in vivo* studies on protective and inhibitory antibodies against *Plasmodium falciparum* in the saimiri monkey, *J. Immunol.*, 132(1), 432, 1984.

46. **Shear, H. L., Nussenzweig, R. S., and Bianco, C.,** Immune phagocytosis in murine malaria, *J. Exp. Med.*, 149, 1288, 1979.

47. **Brown, K. M. and Kreier, J. P.,** *Plasmodium berghei* malaria: blockage by immune complexes of macrophage receptors for opsonized plasmodia, *Infect. Immun.*, 37(3), 1227, 1982.

48. **Hunter, K. W., Winkelstein, J. A., and Simpson, T. W.,** Serum opsonic activity in rodent malaria: functional and immunochemical characteristics *in vitro, J. Immunol.*, 123(6), 2582, 1979.

49. **Alder, J. D. and Kreier, J. P.,** The effect of immune serum on the infectivity of sonically damaged *Plasmodium berghei* infected erythrocytes, *Ohio J. Sci.*, 85(3), 101, 1985.

50. **Jain, S. and Vianyak, V. K.,** Phagocytosis of *Plasmodium chabaudi* modulation by immune serum and antigen *in vitro, J. Hyg. Epidemiol. Microbiol. Immunol.*, 30(2), 207, 1986.

51. **Druilhe, P. and Khusmith, S.,** Epidemiological correlation between levels of antibodies promoting merozoite phagocytosis of *Plasmodium falciparum* and malaria-immune status, *Infect. Immun.*, 55(4), 888, 1987.

52. **Langreth, S. G. and Reese, R. T.,** Antigenicity of the infected-erythrocyte and merozoite surfaces in *falciparum* malaria, *J. Exp. Med.*, 150, 1241, 1979.

53. **Kilejian, A.,** Characterization of a protein correlated with the production of knob-like protrusions on membranes of erythrocytes infected with *Plasmodium falciparum*, *Proc. Natl. Acad. Sci. U.S.A.*, 76(9), 4650, 1979.

54. **Howard, R. J.,** Alterations in the surface membrane of red blood cells during malaria, *Immunol. Rev.*, 61, 67, 1982.

55. **Taylor, D. W., Parra, M., Chapman, G. B., Stearns, M. E., Rener, J., Aikawa, M., Uni, S., Aley, S. B., Panton, L. J., and Howard, R. J.,** Localization of *Plasmodium falciparum* histidine-rich protein 1 in the erythrocyte skeleton under knobs, *Mol. Biochem. Parasitol.*, 25, 165, 1987.

56. **Culvenor, J. G., Langford, C. J., Crewther, P. E., Saint, R. B., Coppel, R. L., Kemp, D. J., Anders, R. F., and Brown, G.,** *Plasmodium falciparum:* identification and localization of a knob protein antigen expressed by a cDNA clone, *Exp. Parasitol.*, 63, 58, 1987.

57. **Schmidt, J. A., Udeinya, I. J., Leech, J. H., Hay, R. J., Aikawa, M., Barnwell, J., Green, I., and Miller, L. H.,** *Plasmodium falciparum* malaria: an amelanotic melanoma cell line bears receptors for the knob ligand on infected erythrocytes, *J. Clin. Invest.*, 70, 379, 1982.

58. **Udeinya, I. J., Miller, L. H., McGregor, I. A., and Jensen, J. B.,** *Plasmodium falciparum* strain-specific antibody blocks binding of infected erythrocytes to amelanotic melanoma cells, *Nature*, 303, 429, 1983.

59. **David, P. H., Hommel, M., Miller, L. H., Udeinya, I. J., and Oligino, L. D.,** Parasite sequestration in *Plasmodium falciparum* malaria: spleen and antibody modulation of cytoadherence of infected erythrocytes, *Proc. Natl. Acad. Sci. U.S.A.*, 80, 5075, 1983.

60. **Wahlin, B., Wahlgren, M., Perlmann, H., Berzins, K., Bjorkman, A., Patarroyo, M. E., and Perlmann, P.,** Human antibodies to a M_r 155,000 *Plasmodium falciparum* antigen efficiently inhibit merozoite invasion, *Proc. Natl. Acad. Sci. U.S.A.*, 81, 7912, 1984.

61. **Collins, W. E., Anders, R. F., Pappaioanou, M., Campbell, G. H., Brown, G. V., Kemp, D. J., Coppel, R. L., Skinner, J. C., Andrysiak, P. M., Favaloro, J. M., Corcoran, L. M., Broderson, J. R., Mitchell, G. F., and Campbell, C. C.,** Immunization of *Aotus* monkeys with recombinant proteins of an erythrocyte surface antigen of *Plasmodium falciparum*, *Nature*, 323, 259, 1986.

62. **Barker, L. R. and Powers, K. G.,** Impairment of antibody response and recovery in malarial rodents by antilymphocyte serum, *Nature*, 229, 429, 1971.

63. **Jayawardena, A. N., Murphy, D. B., Janeway, C. A., and Gershon, R. K.,** T cell-mediated immunity in malaria. I. The Ly phenotype of T cells mediating resistance to *Plasmodium yoelii, J. Immunol.*, 129(1), 377, 1982.

64. **Jayawardena, A. N., Targett, G. A. T., Carter, R. L., Leuchars, E., and Davies, A. J. S.,** The immunological response of CBA mice to *P. yoelii*. I. General characteristics, the effects of T-cell deprivation and reconstitution with thymus grafts, *Immunology*, 32, 849, 1977.

65. **Taylor, D. W., Bever, C. T., Rollwagen, F. M., Evans, C. B., and Asofsky, R.,** The rodent malaria parasite *Plasmodium yoelii* lacks both types 1 and 2 T-independent antigens, *J. Immunol.*, 128(4), 1854, 1982.

66. **Brinkmann, V., Kaufman, S. H., and Simon, M. M.,** T cell-mediated immune response to murine malaria: differential effects of antigen-specific Lyt T cell subsets in recovery from *Plasmodium yoelii* infection in normal and T cell-deficient mice, *Infect. Immun.*, 47, 737, 1985.

67. **Mogil, R. J., Patton, C. L., and Green, D. R.,** Cellular subsets involved in cell-mediated immunity to murine *Plasmodium yoelii* 17X malaria, *J. Immunol.*, 138(6), 1933, 1987.

68. **Hermann, R.,** *Plasmodium chabaudi:* host lymphocyte-macrophage interaction *in vitro, Exp. Parasitol.*, 42, 211, 1977.

69. **Brown, I. N., Watson, S. R., and Sljivic, V. S.,** Antibody response *in vitro* of spleen cells from *Plasmodium yoelii-* infected mice, *Infect. Immun.,* 16(2), 456, 1977.
70. **Loose, L. D. and DiLuzio, N. R.,** A temporal relationship between reticuloendothelial system phagocytic alterations and antibody responses in mice infected with *Plasmodium berghei* (NYU-2 strain), *Am. J. Trop. Med. Hyg.,* 25, 221, 1976.
71. **Lelchuk, R., Rose, E., and Playfair, J. H. L.,** Changes in the capacity of macrophages and T cells to produce interleukins during murine malaria infection, *Cell. Immunol.,* 84, 253, 1984.
72. **Biozzi, G.,** Correlation between genetic regulation of immune responsiveness and host defence against infections and tumours, *Eur. J. Clin. Invest.,* 12, 373, 1982.
73. **Biozzi, G., Mouton, D., Stiffel, C., and Bouthillier, Y.,** A major role of macrophages in quantitative genetic regulation of immunoresponsiveness and antiinfectious immunity, *Adv. Immunol.,* 36, 189, 1984.
74. **Singer, A. and Hodes, R. J.,** Mechanisms of T cell-B cell interaction, *Ann. Rev. Immunol.,* 1, 211, 1983.
75. **Hunter, K. W., Finkelman, F. D., Strickland, G. T., Sayles, P. C., and Scher, I.,** Defective resistance to *Plasmodium yoelii* in CBA/N mice, *J. Immunol.,* 123(1), 133, 1979.
76. **Jayawardena, A. N. and Kemp, J. D.,** Immunity to *Plasmodium yoelii* and *Babesia microti:* modulation by the CBA/N X-chromosome, *Bull. W.H.O.,* 57(Suppl. 1), 155, 1979.
77. **Jayawardena, A. N., Janeway, C. A., and Kemp, J. D.,** Experimental malaria in the CBA/N mouse, *J. Immunol.,* 123(6), 2532, 1979.
78. **Tobie, J. E., Abele, D. C., Wolff, S. M., Contacos, P. G., and Evans, C. B.,** Serum immunoglobulin levels in human malaria and their relationship to antibody production, *J. Immunol.,* 97(4), 498, 1966.
79. **Tobie, J. E., Abele, D. C., Hill, G. J., Contacos, P. G., and Evans, C. B.,** Fluorescent antibody studies on the immune response in sporozoite-induced and blood-induced vivax malaria and the relationship of antibody production to parasitemia, *Am. J. Trop. Med. Hyg.,* 15(5), 676, 1966.
80. **Kuvin, S. F., Tobie, J. E., Evans, C. B., Coatney, G. R., and Contacos, P. G.,** Fluorescent antibody studies on the course of antibody production and serum gamma globulin levels in normal volunteers infected with human and simian malaria, *Am. J. Trop. Med. Hyg.,* 11(4), 429, 1962.
81. **Collins, W. E., Contacos, P. G., Skinner, J. C., Harrison, A. J., and Gell, L. S.,** Patterns of antibody and serum proteins in experimentally induced human malaria, *Trans. R. Soc. Trop. Med. Hyg.,* 65(1), 43, 1971.
82. **Langhorne, J., Evans, C. B., Asofsky, R., and Taylor, D. W.,** Immunoglobulin isotype distribution of malaria-specific antibodies produced during infection with *Plasmodium chabaudi adami* and *Plasmodium yoelii, Cell. Immunol.,* 87, 452, 1984.
83. **Wassom, D. L., Johnson, B. E., and Sayles, P. C.,** A filter immunoplaque assay for quantitation of parasite-specific antibody producing cells, *Parasitol. Today,* 2, 225, 1986.
84. **Ada, G. L. and Byrt, P.,** Specific inactivation of antigen-reactive cells with ^{125}I-labelled antigen, *Nature,* 222, 1291, 1969.
85. **Wen, L., Hanvanich, M., Werner-Favre, C., Brouwers, N., Perrin, L. H., and Zubler, R. H.,** Limiting dilution assay for human B cells based on their activation by mutant EL4 thymoma cells: total and anti-malaria responser B cell frequencies, *Eur. J. Immunol.,* 17, 887, 1987.
86. **Curtain, C. C., Kidson, C., Champness, D. L., and Gorman, J. G.,** Malaria antibody content of gamma$_2$-7S globulin in tropical populations, *Nature,* 203, 1366, 1964.
87. **Butcher, G. A., Cohen, S., and Garnham, P. C. C.,** Passive immunity in *Plasmodium knowlesi* malaria, *Trans. R. Soc. Trop. Med. Hyg.,* 64(6), 850, 1970.
88. **Finerty, J. F., Toby, J. E., and Evans, C. B.,** Antibody and immunoglobulin synthesis in germfree and conventional mice infected with *Plasmodium berghei, Am. J. Trop. Med. Hyg.,* 21, 499, 1972.
89. **Abele, D. C., Tobie, J. E., Contacos, P. G., and Evans, C. B.,** Alterations in serum proteins and 19S antibody production during the course of induced malarial infections in man, *Am. J. Trop. Med. Hyg.,* 14, 191, 1965.
90. **Goumard, P., Vu Dac, N., Maurois, P., and Camus, D.,** Influence of malaria on a pre-existing antibody response to heterologous antigens, *Ann. Immunol.,* 133-D, 313, 1982.
91. **Rosenberg, Y. J.,** Autoimmune and polyclonal B cell response during murine malaria, *Nature,* 274, 170, 1978.
92. **Rosenberg, Y. J. and Chiller, J. M.,** Ability of antigen-specific helper cells to effect a class-restricted increase in total Ig-secreting cells in spleens after immunization with the antigen, *J. Exp. Med.,* 150, 517, 1979.
93. **Rosenberg, Y. J.,** The ability of nonspecific T-cell stimulators to induce helper-cell-dependent increases in either polyclonal or isotype-restricted Ig production *in vivo, Cell. Immunol.,* 61, 416, 1981.
94. **Langhorne, J., Kim, K. J., and Asofsky, R.,** Distribution of immunoglobulin isotypes in the nonspecific B-cell response induced by infection with *Plasmodium chabaudi adami* and *Plasmodium yoelii, Cell. Immunol.,* 90, 251, 1985.
95. **Flanaga, P. B., Lima, M. R. D'I., Coutinho, A., and da Silva, L. P.,** Isotypic pattern of polyclonal B cell response during primary infection by *Plasmodium chabaudi* and in immune-protected mice, *Eur. J. Immunol.,* 17, 599, 1987.

96. **Taylor, D. W., Munoz, P. A., Kim, K. J., Evans, C. B., and Asofsky, R.,** *Plasomdium yoelii:* comparison of indirect immunofluorescence and radioimmunoassay for detecting monoclonal antibodies to malaria, *Exp. Parasitol.*, 53, 362, 1981.

97. **Taylor, D. W., Pacheco, E., Evans, C. B., and Asofsky, R.,** Inbred mice infected with *Plasmodium yoelii* differ in their antimalarial immunoglobuiln isotype response, *Parasite Immunol.*, 10, 33, 1988.

98. **Teale, J. M. and Abraham, K. M.,** The regulation of antibody class expression, 8, 122, 1987.

99. **Isakson, P. C., Pure, E., Vitetta, E. S., and Krammer, P. H.,** T cell-derived B cell differentiation factor(s): effect on the isotype switch of murine B cells, *J. Exp. Med.*, 155, 734, 1982.

100. **Snapper, C. M. and Paul, W. E.,** Interferon-γ and B cell stimulatory factor-1 reciprocally regulate Ig isotype production, *Science*, 236, 944, 1987.

101. **Huang, K.-Y., Schultz, W. W., and Gordon, F. B.,** Interferon induced by *Plasmodium berghei, Science*, 162, 123, 1968.

102. **Biggar, R. J., Collins, W. E., and Campbell, C. C.,** The serological infection in urban Ghanaian infants, *Am. J. Trop. Med. Hyg.*, 29(5), 720, 1980.

103. **Gilles, H. M. and McGregor, I. A.,** Studies on the significance of high serum gamma-globulin concentrations in Gambian Africans. I. Gamma-globulin concentrations of Gambian children in the first two years of life, *Ann. Trop. Med. Parasitol.*, 53, 492, 1959.

104. **McGregor, I. A. and Gilles, H. M.,** Studies on the significance of high serum gamma-globulin concentrations in Gambian Africans. II. Gamma-globulin concentrations of Gambian children in the fourth, fifth, and sixth years of life, *Ann. Trop. Med. Parasitol.*, 54(3), 275, 1960.

105. **Wahlgren, M., Perlman, H., Berzins, K., Bjorkman, A., Larsson, A., Ljungstrom, I., Patarroyo, M. E., and Perlmann, P.,** Characterization of the humoral immune response in the *Plasmodium falciparum* malaria. III. Factors influencing the coexpression of antibody isotypes (IgM and IgG-1 to 4), *Clin. Exp. Immunol.*, 63(2), 343, 1986.

106. **Wahlgren, M., Berzins, K., Perlmann, P., and Person, M.,** Characterization of the humoral immune response in *Plasmodium falciparum* malaria. II. IgG subclass levels of anti-*P. falciparum* antibodies in different sera, *Clin. Exp. Immunol.*, 54, 135, 1983.

107. **Mongini, P. K. A., Paul, W. E., and Metcalf, E. S.,** T-cell regulation of immunoglobulin enhancement of the immunoglobulin switch, *J. Exp. Med.*, 155, 884, 1982.

108. **Anders, R. F.,** Multiple cross-reactivities amongst antigens of *Plasmodium falciparum* impair the development of protective immunity against malaria, *Parasit. Immunol.*, 8, 529, 1986.

109. **Dame, J. B., Williams, J. L., McCutchan, T. F., Wever, J. L., Wirtz, R. A., Hockmeyer, W. T., Maloy, W. L., Haynes, J. D., Schneider, I., Roberts, D., Sanders, G. S., Reddy, E. P., Diggs, C. L., and Miller, L. H.,** Structure of the gene encoding the immunodominant surface antigen on the sporozoite of the human malaria parasite *Plasmodium falciparum, Science*, 225, 593, 1984.

110. **Coppel, R. L., Cowman, A. F., Lingelbach, K. R., Brown, G. V., Saint, R. B., Kemp, D. J., and Anders, R. F.,** Isolate-specific S-antigen of *Plasmodium falciparum* contains a repeated sequence of eleven amino acids, *Nature*, 306, 751, 1983.

111. **Cowman, A. F., Coppel, R. L., Saint, R. B., Favaloro, J., Crewther, P. E., Stahl, H.-D., Bianco, A. E., Brown, G. V., Anders, R. F., and Kemp, D. J.,** The ring-infected erythrocyte surface antigen (RESA) polypeptide of *Plasmodium falciparum* contains two separate blocks of tandem repeats encoding antigenic epitopes that are naturally immunogenic in man, *Mol. Biol. Med.*, 2, 207, 1984.

112. **Hamburger, J. and Kreier, J. P.,** Interaction between protective antibodies and malaria parasites (*Plasmodium berghei*): involvement of low avidity antibodies, *Troppemed. Parasit.*, 27(3), 385, 1976.

113. **Hamburger, J. and Kreier, J. P.,** *Plasmodium berghei:* use of free blood stage parasites to demonstrate protective humoral activity in serum of recovered rats, *Exp. Parasitol.*, 40, 158, 1976.

114. **Steward, M. W. and Voller, A.,** The effect of malaria on the relative affinity of mouse antiprotein antibody, *Br. J. Exp. Pathol.*, 54, 198, 1973.

115. **Wyler, D. J. and Oppenheim, J. J.,** Lymphocyte transformation in human *Plasmodium falciparum* malaria, *J. Immunol.*, 113(2), 449, 1974.

116. **Lustig, H. J., Nussenzweig, V., and Nussenzweig, R. S.,** Erythrocyte membrane-associated immunoglobulins during malaria infection of mice, *J. Immunol.*, 119(1), 210, 1977.

117. **Ronai, Z. and Sulitzeanu, D.,** Anti-red blood cell autoantibodies induced in rat by *Plasmodium berghei* infection bind to a cross-reacting determinant present also in other cell types, *Clin. Immunol. Immunopathol.*, 26, 327, 1983.

118. **Soni, J. L. and Cox, H. W.,** Pathogenesis of acute avian malaria. II. Anemia mediated by a cold-active autohemagglutinin from the blood of chickens with acute *Plasmodium gallinaceum* infection, *Am. J. Trop. Med. Hyg.*, 24, 206, 1975.

119. **Facer, C. A., Bray, R. S., and Brown, J.,** Direct Coombs antiglobulin reactions in Gambian children with *Plasmodium falciparum* malaria. I. Incidence and class specificity, *Clin. Exp. Immunol.*, 35, 119, 1979.

120. **Facer, C. A., Bray, R. S., and Brown, J.,** Direct Coombs antiglobulin reactions in Gambian children with *Plasmodium falciparum* malaria. II. Specificity of erythrocyte-bound IgG, *Clin. Exp. Immunol.,* 39, 279, 1980.

121. **Facer, C. A., Bray, R. S., and Brown, J.,** Direct antiglobulin reactions in Gambian children with *P. falciparum* malaria. III. Expression of IgG subclass determinants and genetic markers and association with anaemia, *Clin. Exp. Immunol.,* 41, 81, 1980.

122. **Musoke, A. J., Cox, H. W., and Williams, J. F.,** Experimental infection with *Plasmodium chabaudi* in rats: antigen and antibody associated with anemia and glomerulonephritis of acute infection, *J. Parasitol.,* 63(6), 1081, 1977.

123. **Wells, R. A., Pavanand, K., Zolyomi, S., Permpanich, B., and MacDermott, R. P.,** Anti-lymphocytotoxic antibodies in sera of Thai adults infected with *Plasmodium falciparum* or *Plasmodium vivax*, *Clin. Exp. Immunol.,* 39, 663, 1980.

124. **Gilbreath, M. J., Pavanand, K., MacDermott, R. P., Wells, R. A., and Ussery, M. A.,** Characterization of cold reactive lymphocytotoxic antibodies in malaria, *Clin. Exp. Immunol.,* 51, 232, 1983.

125. **De Souza, J. B. and Playfair, J. H.,** Anti-lymphocyte antibodies in lethal mouse malaria. II. Induction of an autoantibody specific suppressor T cell by non-lethal *P. yoelii*, *Clin. Exp. Immunol.,* 54, 110, 1983.

126. **Packer, B. J. and Kreier, J. P.,** *Plasmodium berghei* malaria: effects of acute-phase serum and erythrocyte-bound immunoglobulins on erythrophagocytosis by rat peritoneal macrophages, *Infect. Immun.,* 51(1), 141, 1986.

127. **Nussenzweig, R. S., Cochrane, A. H., and Lustig, H. J.,** Immunological responses, in *Rodent Malaria,* Killick-Kendrick, R. and Peters, W., Eds., Academic Press, New York, 1978, chap. 6.

128. **Spitalny, G. L. and Nussenzweig, R. S.,** *Plasmodium berghei:* relationship between protective immunity and anti-sporozoite (CSP) antibody in mice, *Exp. Parasitol.,* 33, 168, 1973.

129. **Hansen, R., deSilva, S., and Strickland, G. T.,** Antisporozoite antibodies in mice immunized with irradiation-attenuated *Plasmodium berghei* sporozoites, *Trans. R. Soc. Trop. Med. Hyg.,* 73(5), 574, 1979.

130. **Beaudoin, R. L., Strome, C. P., Mitchell, F., and Tubergen, T. A.,** *Plasmodium berghei:* immunization of mice against the ANKA strain using the unaltered sporozoite as an antigen, *Exp. Parasitol.,* 42, 1, 1977.

131. **Nussenzweig, R. S., Vanderberg, J. P., and Most, H.,** Protective immunity produced by the injection of X-irradiated sporozoites of *Plasmodium berghei.* IV. Dose response, specificity and humoral immunity, *Mil. Med.,* 134, 1176, 1969.

132. **Nussenzweig, R. S., Vanderberg, J. P., Sanabria, Y., and Most, H.,** *Plasmodium berghei:* accelerated clearance of sporozoites from blood as part of immune-mechanism in mice, *Exp. Parasitol.,* 31, 88, 1972.

133. **Verhave, J. P., Meuwissen, J. H. W. Th., and Golenser, J.,** Cell-mediated reactions and protection after immunization with sporozoites, *Israel J. Med. Sci.,* 14(5), 611, 1978.

134. **Spitalny, G. L., Rivera-Ortiz, C.-I., and Nussenzweig, R. S.,** *Plasmodium berghei:* the spleen in sporozoite-induced immunity to mouse malaria, *Exp. Parasitol.,* 40, 179, 1976.

135. **Yoshida, N., Potocnjak, P., Nussenzweig, V., and Nussenzweig, R. S.,** Biosynthesis of Pb44, the protective antigen of sporozoites of *Plasmodium berghei, J. Exp. Med.,* 154, 1225, 1981.

136. **Potocnjak, P., Yoshida, N., Nussenzweig, R. S., and Nussenzweig, V.,** Monovalent fragments (Fab) of monoclonal antibodies to a sporozoite surface antigen (Pb44) protect mice against malarial infection, *J. Exp. Med.,* 151, 1504, 1980.

137. **Zavala, F., Cochrane, A. H., Nardin, E. H., Nussenzweig, R. S., and Nussenzweig, V.,** Circumsporozoite proteins of malaria parasites contain a single immunodominant region with two or more identical epitopes, *J. Exp. Med.,* 157, 1947, 1983.

138. **Aikawa, M., Yoshida, N., Nussenzweig, R. S., and Nussenzweig, V.,** The protective antigen of malarial sporozoites (*Plasmodiu berghei*) is a differentiation antigen, *J. Immunol.,* 126(6), 2494, 1981.

139. **Hollingdale, M. R., Zavala, F., Nussenzweig, R. S., and Nussenzweig, V.,** Antibodies to the protective antigen of *Plasmodium berghei* sporozoites prevent entry into cultured cells, *J. Immunol.,* 128(4), 1929, 1982.

140. **Chen, D. H., Tigelaar, R. E., and Weinbaum, F. I.,** Immunity to sporozoite-induced malaria infection in mice. I. The effect of immunization of T and B cell-deficient mice, *J. Immunol.,* 118(4), 1322, 1977.

141. **Hoffman, S. L., Oster, C. N., Plowe, C. V., Woollett, G. R., Beier, J. C., Chulay, J. D., Wirtz, R. A., Hollingdale, M. R., and Mugambi, M.,** Naturally acquired antibodies to sporozoites do not prevent malaria: vaccine development implications, *Science,* 237, 639, 1987.

142. **Tapchaisri, P., Asavanich, A., Limsuwan, S., Tharavanij, S., and Harinasuta, K. T.,** Antibodies against malaria sporozoites in patients with acute uncomplicated malaria and patients with cerebral malaria, *Am. J. Trop. Med. Hyg.,* 34(5), 831, 1985.

143. **Hoffman, S. L., Wistar, R., Ballou, W. R., Hollingdale, M. R., Wirtz, R. A., Schneider, I., Marwoto, H. A., and Hockmeyer, W. T.,** Immunity to malaria and naturally acquired antibodies to the circumsporozoite protein of *Plasmodium falciparum, N. Engl. J. Med.,* 315(10), 601, 1986.

144. **Nardin, E. H., Nussenzweig, R. S., Bryan, J. H., and McGregor, I. A.,** Congenital transfer of antibodies against malarial sporozoites detected in Gambian infants, *Am. J. Trop. Med. Hyg.,* 30(6), 1159, 1981.

145. **Webster, H. K., Boudreau, E. F., Pang, L. W., Permpanich, B., Sookto, P., and Wirtz, R. A.,** Development of immunity in natural *Plasmodium falciparum* malaria: antibodies to the falciparum sporozoite vaccine 1 antigen (R32tet32), *J. Clin. Microbiol.*, 25(6), 1002, 1987.

146. **Schofield, L., Villaquiran, J., Ferreira, A., Schellekens, H., Nussenzweig, R., and Nussenzweig, V.,** Gamma interferon, CD8$^+$ T cells and antibodies required for immunity to malaria sporozoites, *Nature*, 330, 664, 1987.

147. **Charoenvit, Y., Leef, M. F., Yuan, L. F., Sedegah, M., and Beaudoin, R. L.,** Characterization of *Plasmodium yoelii* monoclonal antibodies directed against stage-specific sporozoite antigens, *Infect. Immun.*, 55(3), 604, 1987.

148. **Spitalny, G. L., Verhave, J. P., Meuwissen, J. H. E. Th., and Nussenzweig, R. S.,** *Plasmodium berghei:* T cell dependence of sporozoite-induced immunity in rodents, *Exp. Parasitol.*, 42, 73, 1977.

149. **Good, M. F., Berzofsky, J. A., Maloy, W. L., Hayashi, Y., Fujii, N., Hockmeyer, W. T., and Miller, L. H.,** Genetic control of the immune response in mice to a *Plasmodium falciparum* vaccine: widespread nonresponsiveness to single malaria T epitope, *J. Exp. Med.*, 164, 655, 1986.

150. **Grau, G. E., Del Giudice, G., and Lambert, P.-H.,** Host immune response and pathological expression in malaria: possible implications for malaria vaccines, *Parasitology*, 94, S123, 1987.

151. **Del Giudice, G., Engers, H. D., Tougne, C., Biro, S. S., Weiss, N., Verdini, A. S., Pessi, A., Degremont, A. A., Freyvogel, T. A., Lambert, P.-H., and Tanner, M.,** Antibodies to the repetitive epitope of *Plasmodium falciparum* circumsporozoite protein in rural Tanzanian community: a longitudinal study of 132 children. *Am. J. Trop. Med. Hyg.*, 36, 203, 1987.

152. **Gwadz, R. W.,** Malaria: successful immunization against the sexual stages of *Plasmodium gallinaceum*, *Science*, 193, 1150, 1976.

153. **Carter, R. and Chen, D. H.,** Malaria transmission blocked by immunisation with gametes of the malaria parasite, *Nature*, 263(5572), 57, 1976.

154. **Mendis, K. N. and Targett, G. A. T.,** Immunization to produce a transmission-blocking immunity to *Plasmodium yoelii* malaria infections, *Trans. R. Soc. Trop. Med. Hyg.*, 75, 158, 1981.

155. **Rener, J., Graves, P. M., Carter, R., Williams, J., and Burkot, T. R.,** Target antigens of transmission blocking immunity on gametes of *Plasmodium falciparum*, *J. Exp. Med.*, 158, 976, 1983.

156. **Mendis, K. N., Munesinghe, Y. D., de Silva, Y. N. Y., Keragalla, I., and Carter, R.,** Malaria transmission-blocking immunity induced by natural infections of *Plasmodium vivax* in humans, *Infect. Immun.*, 55(2), 369, 1987.

157. **Kaushal, D. C., Carter, R., Rener, J., Grotendorst, C. A., Miller, L. H., and Howard, R. J.,** Monoclonal antibodies against surface determinants on gametes of *Plasmodium gallinaceum* block transmission of malaria parasites to mosquitoes, *J. Immunol.*, 131, 2557, 1983.

158. **Vermeulen, A. N., Ponnudurai, T., Beckers, P. J. A., Verhave, J. P., Smits, M. A., and Meuwissen, J. H. W. Th.,** Sequential expression of antigens on sexual stages of *Plasmodium falciparum* accessible to transmission-blocking antibodies in the mosquito, *J. Exp. Med.*, 162, 1460, 1985.

159. **Vermeulen, A. N., van Deursen, J., Brankenhoff, R. H., Lensen, T. H. W., Ponnudurai, T., and Meuwissen, J. H. W. Th.,** Characterization of *Plasmodium falciparum* sexual stage antigens and their biosynthesis in synchronised gametocyte cultures, *Mol. Biochem. Parasitol.*, 20, 155, 1986.

160. **Ponnudurai, T., van Gemert, G. J., Bensink, T., Lensen, A. H. W., and Meuwissen, J. H. W. Th.,** Transmission blockade of *Plasmodium falciparum*: its variability with gametocyte numbers and concentration of antibody, *Trans. R. Soc. Trop. Med. Hyg.*, 81, 491, 1987.

Chapter 2

ROLE OF CELL-MEDIATED IMMUNITY IN RESISTANCE TO MALARIA

Johanne Melancon-Kaplan and William P. Weidanz

TABLE OF CONTENTS

I. INTRODUCTION

Immunity to malaria is established under natural conditions but requires repeated infections, takes years to develop, and is not absolute in that parasitemia may persist in the absence of clinical disease. In addition, immunity to malaria is short lived in the absence of repeated infection as evidenced by outbreaks of epidemic malaria following the termination of control programs and the susceptibility of previously immune individuals who return to areas of malaria transmission after spending periods of less than 1 year away from a malarious area.[1,2] Similar conclusions have been derived from experimental infections in humans. Observations that resistance to plasmodia could be achieved by the passive transfer of immune sera provided further evidence for the role of acquired immunity in malaria and helped to establish the concept that immunity to malaria is mediated by antibodies.[3]

As the worldwide malaria situation continued to deteriorate, the search for better ways of controlling malaria was intensified and the idea that this could be achieved by immunization gained acceptance and support within the scientific community.[4,5] At the same time, powerful new technologies were being developed and, as they became available, were applied to the problem of malaria in hopes of identifying antigens which could be used to stimulate the production of protective antibodies. Subsequently, putative protective antigens were identified in sporozoites, merozoites, gametocytes, and in the membranes of infected erythrocytes. The genes for certain of these stage-specific antigens have been cloned, sequenced, and their corresponding peptides have been expressed by means of recombinant DNA techniques or have been synthesized in the laboratory. Several of these antigens have already been tested in humans[6,7] and, while the results of these studies were less favorable than originally anticipated, they serve as prototypes for ongoing and future studies aimed at producing an efficacious malaria vaccine.

We now know more about the molecular biology and biochemistry of plasmodia than ever before. Moreover, we are learning new ways of enhancing immunogenicity and identifying new targets for chemotherapeutic intervention. Despite these remarkable advances, the identification and characterization of particular host responses which suppress parasite growth or actually kill plasmodia remain for the most part to be accomplished.[8] In fact, we do not know with certainty how plasmodia establish stable relationships with their hosts nor how immune responses clear plasmodia from host tissues or prevent uncontrolled parasite growth. What is becoming apparent is that immunity to malaria is dependent on antibody- as well as cell-mediated mechanisms with antigen-specific T cells playing an essential role in both.[5,9] It seems likely that the characterization of these cells and their functions as well as the identification of antigenic epitopes which activate them will prove useful in the rational design of candidate malarial vaccines. While extensive literature exists regarding the role of protective antibodies in malaria,[10,11] considerably less is known about the function of cell-mediated immunity in resisting this disease.

The purpose of this report is to review the evidence for the role of cell-mediated immunity (CMI) in malaria.

II. THYMUS DEPENDENCE OF IMMUNITY TO MALARIA

In 1968 Brown et al.[12] reported that *Plasmodium berghei* infections in neonatally thymectomized rats were more severe than those observed in euthymic animals. The thymic dependence of resistance to malaria was subsequently confirmed by others using both rat and mouse models of rodent malaria.[13-15] In addition, congenitally athymic nude mice succumbed to infection when injected with normally avirulent murine plasmodia.[16] Acute infections terminated in such mice by drug therapy were observed to recrudesce, an event which did not occur in thymic-grafted nude mice.[17] Recrudescence also occurred in chronically infected rats following treatment with antithymocyte sera.[18] Together these data demonstrated an essential role for the thymus in the development of immunity to malaria but did not explain the mechanisms involved. Brown[18] originally postulated that thymectomy depleted helper T cells required for the production of protective antibodies by B cells to sequentially appearing variants of the parasite. Allison and Clark[19] suggested that T cell-dependent cell-mediated immune mechanisms were suppressed by these procedures. It now seems likely that both explanations have merit.

III. EVIDENCE FOR THE ROLE OF CELL-MEDIATED IMMUNITY IN MALARIA

The protective role of antibody against erythrocytic malaria has recently been discussed by others[10,11] and is reviewed elsewhere in this volume. Other evidence suggests that immunity to malaria is mediated by mechanisms which act in concert with or independently of protective antibodies.[20] For example, no serological tests are currently available to correlate antibody titers with functional immunity[9] and it is well established that passive immunization with "immune" sera is difficult to achieve, requires the transfer of large volumes of serum, and is complicated further by the use of thymectomized or splenectomized recipients.[21-24] In murine models, immunity to malaria has been associated with the development of delayed-type hypersensitivity (DTH).[25,26]

In addition, antibody-independent resistance to assorted hemoprotozoa has been achieved by the injection of various immunomodulating agents.[19] The most compelling evidence that nonantibody T cell-dependent immune mechanisms can mediate resistance to malaria has been obtained from studies utilizing B cell-deficient hosts which resolved acute infections 9with blood-stage parasites spontaneously[27] or resisted reinfection following chemotherapy.[28-30] The finding of Chen et al.[31] that B cell-deficient mice immunized with irradiated *P. berghei* sporozoites were resistant to challenge with viable sporozoites demonstrated that resistance to this stage of the parasite could be mediated by T cell-dependent immune mechanisms as well. The results of adoptive transfer studies employing immune T cells, antigen-specific T cell lines, and T cell clones provide definitive evidence in support of this concept (see below).

A. STIMULATION OF NONSPECIFIC IMMUNITY TO HEMOPROTOZOA

A variety of immunomodulating agents (e.g., *Corynebacterium parvum*, BCG) have been used to activate resistance mechansims capable of protecting animals against experimental infection with hemoprotozoa.[19,20,32] The protection achieved has ranged from a brief delay in the onset of patent parasitemia to complete suppression of disease.[33] Also, the degree of protection achieved with particular agents was dependent upon the infecting stage of the parasite being tested[34] and was expressed selectively against different species of plasmodia.[35] The host cell preference of the parasite appeared to influence the potency of nonspecific immunity activated by a particular agent with parasites of mature erythrocytes being the most susceptible.[36] In general, nonspecific immunity resulting from the injection of immunomodulating agents was relatively short lived, lasting for periods of days to months, and lacked a memory component.[37] Where tested, nonspecific immunity against hemoprotozoa was not attributed to the presence of protective antibody.[38,39] The significance of these observations is that plasmodia were eliminated from host tissues through the stimulation of host responses which either killed the parasites or inhibited their growth in the absence of detectable antibodies. Whether the mechanisms involved were activated during the course of natural infection or were peculiar to the inducing agent has been difficult to decipher.

B. IMMUNITY TO MALARIA IN B CELL-DEFICIENT HOSTS

The demonstration that immunity to malaria can be mediated by T cell-dependent nonantibody mechanisms of immunity has been obtained from experiments utilizing B cell-deficient animals and assorted species of plasmodia.[40] Chickens rendered B cell-deficient by means of surgical or chemical bursectomy developed more severe malaria than immunologically intact birds when infected with *P. lophurae* or *P. gallinaceum*, the latter being fatal for B cell-deficient chickens.[30,41] These findings indicated an essential role for B cells and their antibodies in immunity. They also suggested that acute *P. lophurae* infections could be resolved by nonantibody mediated mechanisms. However, since the bursectomy procedure employed to make birds B cell-deficient failed to completely suppress immunoglobulin synthesis and antibody formation, this issue remains in doubt.

Increased severity of acute malarial infection was similarly reported in mice made B cell-deficient by means of lifelong treatment with anti-mouse u chain sera.[42,43] These defective mice died when infected with blood-stage parasites of the normally avirulent 17X strain of *P. yoelii*. Severe *P. yoelii* malaria of prolonged duration was also reported in CBA/N mice which lack the Lyb5$^+$ subset of B lymphocytes.[44,45] While these data demonstrated that B cells are required to survive acute infections with blood-stage parasites, further studies in B cell-deficient hosts indicated that additional mechanisms of immunity independent of antibodies played a significant role in resistance to malaria. When B cell-deficient hosts were cured of their otherwise lethal malaria by means of chemotherapy, they subsequently developed low-grade malaria of prolonged duration and resisted challenge with homologous parasites.[43] This immunity to reinfection appeared to be T cell-dependent since fatal malaria occurred in irradiated-thymectomized chickens and in nude mice following chemotherapeutic treatment of their acute infections.[17,46]

In certain instances, B cell-deficient hosts resolved acute malarial infections spontaneously in the absence of chemotherapy. This was first observed in B cell-deficient mice infected with *P. chabaudi adami*.[27] More recently, B cell-deficient mice have been shown to resolve acute infections with *P. vinckei petteri, P. chabaudi chabaudi*, and *Babesia microti*.[47] Athymic nude mice died when infected with these parasites. These results suggest that nonantibody T cell-dependent cellular immune mechanisms can function to prevent reinfection or to suppress acute disease. T cell-dependent CMI shows limited specificity in that B cell-deficient mice immune to *P. chabaudi adami* resisted infection with the closely related *P. vinckei vinckei* but died with fulminating malaria when infected with the distantly related *P. yoelii*.[28] Conversely, B cell-deficient mice immunized against *P. yoelii* resisted challenge with *P. chabaudi adami* but became infected when challenged with *P. vinckei*.

Taken together these findings have certain implications; they suggest that immunity to blood-stage murine plasmodia can be mediated by different immune mechanisms depending upon the infecting species of hemoprotozoa, i.e., both antibodies and CMI are required to suppress *P. yoelii* whereas T cell-dependent CMI, in the absence of antibodies, is sufficient for resistance to *P. chabaudi adami*. Following the resolution of acute malaria, nonantibody mechanisms of immunity suffice to keep endogenous parasites at low levels. This premunition required the continued presence of parasites in B cell-deficient hosts. While these studies were conducted with blood-stage parasites, Chen et al.,[31] as indicated above, demonstrated an essential role for T cell-dependent CMI to sporozoites. B cell- deficient mice, in contrast to nude mice, were protected against challenge infection with *P. berghei* sporozoites following immunization with irradiated sporozoites.

C. MALARIA INDUCES T CELL ACTIVATION
1. Proliferation Studies

Plasmodial antigens have been shown to activate T cells in various ways as indicated by T cell proliferation both *in vivo* and *in vitro*, the induction and expression of DTH, the production of lymphokines, and the stimulation of helper cell function necessary for antibody formation. Jayawardena et al.[48] utilized the T6 chromosome marker in radiation chimeras to show that splenic T cells proliferated in response to infection with avirulent *P. yoelii*. Splenic T cell proliferation was also evident in *P. yoelii*-immunized mice following challenge with homologous parasites. While T cell proliferation occurred in mice infected with the lethal *P. berghei*, it was diminished in comparison to that seen in nonlethal malaria and waned as the infection progressed. Similar findings were reported from studies in which T cell proliferation was assessed by enumerating Thy-1[+] cells.[49] Jayawardena et al.[48] concluded that T cell activation *in vivo* correlated with the development of protective immunity.

Subsequently it was observed that T cells from immune hosts proliferated *in vitro* when stimulated with crude preparations of malarial antigens.[50,51] T cells from nonimmune hosts also proliferated in response to these same antigens, although to a lesser degree, suggesting that such antigenic preparations contained mitogenic factors. T cell proliferation was diminished during malaria due to the activation of antigen-specific T suppressor cells.[52,54] Recent studies with murine and human T cell lines and clones reactive with malarial antigens demonstrated that T cells proliferated in an antigen-specific and MHC-restricted manner.[55,56] Although relatively few proliferation studies have employed well-characterized plasmodial antigens, we anticipate that more will be learned about the fine specificity of T cell activation as defined antigens become available through recombinant DNA techniques or are synthesized in the laboratory. It should be noted that antigenic preparations from all life stages of plasmodia have been shown to cause T cell proliferation.

2. Delayed-Type Hypersensitivity

Antigen-specific T cell activation in response to malarial antigens has been assessed by the sensitization for and expression of DTH. Indeed, DTH has been elicited during the course of malaria as well as after recovery and in animals immunized with plasmodial antigens.[9,10] Controversy exists regarding the significance of DTH to plasmodial antigens, that is, is DTH a correlate of CMI to malaria? Finerty and Krehl[57] reported that mice pretreated with cyclophosphamide resisted lethality when infected with the virulent 17XL strain of *P. yoelii* and concurrently developed significant DTH to erythrocytic stage parasites. Mice not treated with cyclophosphamide failed to develop DTH and died. The positive outcome of infection was associated with the development of DTH which was thought to be suppressed by cyclophosphamide-sensitive T suppressor cells activated by virulent *P. yoelii*. Other investigators[58] were unable to correlate DTH with the acquisition of immunity during malarial infections. This may be due to a lack of correlation between DTH and CMI or might reflect the fact that the mixtures of malarial antigens which have been used to elicit DTH have lacked sufficient concentrations of protective antigens. In addition, alterations in local inflammatory responses are known to occur during murine malaria and these may influence the development of classical DTH reactions in skin following administration of the eliciting dose of antigen.[59]

Several researchers[60-62] have observed that mice immunized either by *P. yoelii* infection, a crude vaccine consisting of formalin-fixed parasites, or a purified 230-kDa merozoite antigen derived from the same parasite developed significant DTH responses to *P. yoelii* blood-stage antigens. Moreover, the mice resisted challenge infections with viable *P. yoelii*. DTH responses to malarial antigen were specific for the sensitizing plasmodial species, adoptively transferred with spleen cells but not with serum, and mapped to T cells bearing the Lyt1[+],2[-] (probably CD4[+] or helper/inducer T cells) phenotype.[63] The observation by Cottrell et. al.[60] that radiolabeled bone marrow cells homed to the spleens of immunized mice challenged with viable parasites provides an explanation for negative or weak DTH skin tests seen by others during infection. Namely, T cells and monocytes migrating into evolving delayed-type lesions in the spleens and livers of malarious mice might not be available to participate in the development of new lesions induced in the skin by the injection of test antigen.

Recent studies in our laboratory[64] have shown that mice can be sensitized with crude blood-stage antigens of *P. chabaudi adami* injected s.c. DTH was mediated by CD4$^+$ T cells. The same antigens activated CD8$^+$ T cells (cytotoxic/suppressor) when injected by the i.v. route. The function of these antigen-reactive T cell populations remains unknown; however, the development of such model systems may allow for future studies aimed at delineating mechanisms whereby T cell responses to malarial antigens are regulated.

3. Lymphokine Production

The T cell-dependent manifestations of immunity in malaria have been reviewed elsewhere[33] and include splenomegaly, lymphotoxin production, macrophage activation, blood monocyte production, and possibly erythropoiesis. Most of these phenomena can be attributed to the function of lymphokines secreted by activated T cells responding to malarial antigens. Wyler and Gallin[65] identified and partially characterized a mononuclear cell chemotactic factor in extracts of spleen cells derived from malarious mice and monkeys but not from uninfected control animals. Since splenic extracts prepared from infected nude mice lacked this activity, they concluded that the chemotactic factor was produced by T cells or its production was dependent upon T cells. The presence of interferon (IFN) has been reported in the sera of mice and humans with malaria.[66,67] Recent studies demonstrated that T cells from malarious patients and from immune individuals living in endemic areas produced IFN-γ and IL-2 in response to stimulation with homologous antigens.[68] The production of IFN-γ by human T cells was proposed to represent a useful indicator of CMI.

The secretion of IL-2 by concanavalin A (Con A)-stimulated spleen cells of mice infected with *P. yoelii* or *P. berghei* indicated that the ability to produce this lymphokine increased early during infection but then declined.[69] Spleen cells from mice infected with nonlethal *P. yoelii* produced large amounts of IL-2 when stimulated with Con A at the time of recovery. Con A-induced blast cells from spleen cells of malarious mice responded to IL-2 despite the fact that they could not produce this lymphokine. This suggested that malaria produced a defect at the level of IL-2 synthesis. With available technologies it should be possible to define the defect in molecular terms, i.e., the defect may be pre- or post-transcriptional or possibly due to a secretory problem or to the secretion of a nonfunctional peptide. Recent studies by the same authors[70] have revealed the presence of a serum IL-2 inhibitor in euthymic but not athymic mice infected with murine plasmodia. This factor appears to inhibit IL-2 production as well as to block IL-2-dependent T cell proliferation.

4. T Cell Help for Antibody Formation

The activation of helper T cells leads to the production of antibody formation by B lymphocytes. In malaria, T cell-dependent isotypes of antibody have been demonstrated in the sera of animals and humans infected previously with plasmodia.[2,10] Experimentally, murine plasmodia have been used successfully as hapten carriers to stimulate infection-primed T cells to provide help to B cells for the production of hapten-specific antibody.[71] In other experiments, lymphoid cells derived from patients who had experienced *P. falciparum* malaria[72] or from mice immunized with a synthetic *P. vivax* sporozoite peptide produced antibodies in response to immunization *in vitro*.[73] Lymphocytes from unprimed hosts failed to respond. It seems likely that the detectable antibodies were produced by B cells receiving help from T cells activated by plasmodial antigens.

D. ADOPTIVE TRANSFER OF IMMUNITY
1. Transfer of Primary Spleen Cells

In order to study the contribution of T lymphocytes to the immune response against malarial parasites, various investigators have performed transfer experiments in which immune spleen cells or selected spleen cell populations were administered to nonimmune recipient animals. Most studies examined blood-stage *P. yoelii* or *P. chabaudi* infections in mice and *P. berghei* in rats. In every instance, unfractionated immune spleen cells were able to confer protection when transferred into naive animals.[21,74-83] Cell fractionation experiments undertaken to determine the identity of the protective spleen cell population(s) revealed that, in these models, both immune B and T lymphocytes possessed the ability to transfer some level of protection to the recipient[21,63,74,76-82,84,85] and some investigators reported that, when transferred together, the two cell populations synergized to generate significantly enhanced levels of protective activity.[63,77,78] Such results support the hypothesis that immune T cells exert their protective action against malaria by functioning as T helper cells in the production of antibodies by B lymphocytes rather than acting in classical cell-mediated immune responses.[63,77,78,81]

However, Jayawardena et al.[45,63] presented evidence suggesting that, although B cells promoted recovery from an acute *P. yoelii* blood-stage infection, they were not essential for the maintenance of resistance to reinfection which they attributed primarily to cell-mediated mechanisms. This was demonstrated by the observation that B cell-deficient CBA/N mice, which suffered more severe and long-lasting *P. yoelii* infections than normal mice, nevertheless were able to resist reinfection with homologous parasites as well as normal mice.[44,45] Although CBA/N mice were as resistant as normal controls, B cells from these defective animals transferred considerably less immunity to naive recipients than B cells from normal mice,[45] thus indicating that resistance to reinfection resided mostly in a different non-B cell population. These observations are in accordance with the results obtained with agammaglobulinemic chickens and u-suppressed mice.[30,43] Subsequent experiments showed that immune T cells of the Lyt1$^+$ phenotype, as opposed to Lyt2$^+$ T cells, were able to transfer immunity to *P. yoelii* in nonimmune mice.[63,85] Lyt1$^+$ T cells appeared to require interaction with normal host T cells in order to mediate protection against *P. yoelii*[81,85] since adoptive resistance in athymic nude recipients was only observed after prior reconstitution with normal T cells.[85] Fahey and Spitalny[81] suggested that the function of the T cells of the recipient may be to expand the immune response following interaction with the transferred population. More recently, Mogil et al.[84] reported that both Lyt1$^+$ and Lyt2$^+$ immune T cells were able to transfer resistance to *P. yoelii* but that protection was dependent upon the cotransfer of an IJ$^+$, non-T, non-B, immune null cell. These results are similar to the earlier findings of Strickland et al.[75] who reported that a radio-resistant, silica-inactivated, non-T, non-B cell was required along with immune T or B lymphocytes for the optimal expression of transferred immunity to *P. yoelii*. The identity of this null cell remains to be established but it was suggested that it may be a macrophage similar to previously described IJ$^+$ antigen-presenting cells with immunoregulatory activity.[86,87] Mogil et al.[84] tentatively attributed the failure of other investigators to detect protective activity in Lyt2$^+$ immune spleen cell populations to differences in the time at which spleen cells were obtained from donor mice resulting in different activities in the cell populations recovered for transfer.

Since resistance to acute infection with *P. yoelii* and *P. berghei* is antibody dependent,[21,77,78,81] the use of these models in studies of T cell function is likely to obscure the involvement of T lymphocytes in cell-mediated immune mechanisms. In contrast, we have shown that resistance of mice to *P. chabaudi adami* is strictly dependent on cell-mediated immune mechanisms.[27] Therefore, we utilized this model system to investigate the involvement of T lymphocytes in antibody-independent immunity. Cavacini et al.[80] showed that athymic nude mice, which are extremely susceptible to *P. chabaudi adami* infection, were able to control their parasitemia and eliminate the parasite when provided with spleen cells from normal or immune mice. Cell separation indicated that the protective activity of splenocytes resided in the T cell-enriched fraction with the B cell-enriched population being unable to suppress *P. chabaudi adami* infections. Further characterization of the lymphocytes involved revealed that protection was best achieved with T cells of the L3T4$^+$ phenotype.

It is noteworthy that in this study lymphoid cells transferred even in very large numbers failed to suppress the initial appearance of parasites in the blood, thereby suggesting that the grafted cells did not directly kill the parasites but rather functioned as inducers of host effector mechanisms.

Although the studies described above were primarily concerned with blood-stage malaria, adoptive transfer experiments have also been used to study the role of lymphocytes in host immunity to other stages of the parasite. For instance, Verhave et al.[88] reported that anti-*P. berghei* sporozoite immunity could be passively transferred to naive mice with immune spleen cells. Treatment of the immune splenocytes with antitheta and complement abolished their ability to transfer protection thus indicating that T lymphocytes were active in antisporozoite immunity. More recently, Egan et al.[89] investigated the immune response of mice to the *P. berghei* circumsporozoite (CS) protein expressed in *Escherichia coli* and to synthetic peptides from the repeat region of the CS protein conjugated to keyhole limpet hemocyanin. Immunized mice developed high titers of antisporozoite antibodies against the CS protein and their sera were shown to block sporozoite invasion of hepatoma cells *in vitro* and transfer protective activity *in vivo*. Nonetheless, mice that had been immunized with whole irradiated sporozoites displayed far greater resistance to infection than the mice given the recombinant or synthetic peptides even though their serum titers were comparable. This suggested that antibody-independent immune mechanisms were also elicited by sporozoites. In accordance with this proposition, the authors showed that T-enriched, but not B-enriched, spleen cells from sporozoite-immunized mice were able to transfer protection to nonimmune animals while spleen cells from mice given the subunit peptides failed to protect recipients. These results indicate that the role of T cells in resistance to sporozoite infection is not limited to that of a helper function in antibody production but presumably involves the induction of other cellular effector mechanisms.

T lymphocytes have also been implicated in transmission-blocking immunity. Harte et al.[90] reported that vaccination of mice with microgametes of *P. yoelii nigeriensis* successfully blocked transmission of the parasite for at least 12 months despite the fact that serum antibodies to gametocytes were undetectable after 6 months. The passive transfer of immune T cells of the Lyt1$^+$ phenotype reduced transmission of a subsequent infection by 95% with a significant reduction in the number of circulating gametocytes. In contrast, the administration of immune serum was minimally effective but, when given together with immune T cells, showed a strong additive effect and resulted in the complete suppression of the mouse-vector transfer of gametocytes. These results suggest that both humoral and CMI contribute to the control of malaria transmission.

2. Transfer of T Cell Lines and Clones

In recent years, various investigators have reported the development of human or rodent continuous T cell lines[55,91-93] and clones[56,94-99] reactive with plasmodial antigens. Human T cells responding to antigens present in the asexual blood stage[56,94,95,98,99] or gametocytes[97] of *P. falciparum* were found to be present in the peripheral blood of individuals with or without previous exposure to malaria. Antigen-reactive T cells could typically be maintained in culture as lines in the presence of IL-2 and periodic "boosting" with malarial antigen and antigen-presenting cells. The lines could also be cloned by limiting dilution to generate specific T cell clones that were then maintained in the same fashion. The presence of a high frequency of *P. falciparum* antigen-reactive T cells in the blood of donors not previously exposed to the parasite was somewhat disconcerting. It was suggested that this could possibly have resulted from (1) nonspecific stimulation of T lymphocytes by mitogenic components of the parasite, (2) the sensitization of naive T cells during the course of *in vitro* culture with malarial antigen, or (3) the presence of T cells sensitized to common environmental antigens cross-reactive with malarial antigens.[95,97-99] The first hypothesis involving the mitogenicity of parasite antigen preparations appears unlikely since the proliferative response of *P. falciparum*-reactive T cell clones stimulated with antigen required the presence of antigen-presenting cells sharing HLA class II antigens[56,98,99] while PHA, a known T cell mitogen, was able to trigger proliferation of T cell clones in the presence of nonhistocompatible antigen-presenting cells.[98] The specificity of T cell clone responses to antigen preparations from parasitized erythrocytes was also demonstrated by their lack of reactivity to normal red blood cell components.[56,97-99] A relatively large proportion of the T cell clones obtained by different investigators (approximately 50%) appeared to recognize cross-reactive antigens present not only in *P. falciparum* but in species as varied as *P. berghei*[56,95,99] and *P. gallinaceum.*[99] The remainder of the clones appeared to be species specific.[56,95,99] Within the *P. falciparum* species, T lymphocyte clones seemed largely to recognize "constant" antigens present in strains of different geographic origin as 33/34 of the *P. falciparum* specific human T cell clones derived from individuals exposed to malaria in a recent study by Pink et al.[95] responded to lysates of erythrocytes infected with parasites from East or West Africa as well as Asia. Only 1/34 T cell clones distinguished between the different isolates[95] indicating that T cells reactive to polymorphic antigenic determinants of the parasite also develop during the course of the immune response although at a lower frequency. Common T cell-stimulating antigens also appear to be shared by the different life stages of the parasite. *P. falciparum*-specific T cell clones were found by Sinigaglia and Pink[56] to respond to lysates from synchronized cultures containing either immature (ring) or mature (trophozoite and schizont) parasite forms while Good et al.[97] reported that the majority of their gamete-specific T cell clones also reacted with intact gametocytes and asexual-stage parasites. Stage-specific T cell antigens are also present in the parasite since T cell clones could be obtained that responded preferentially[56] or strictly[97] to a specific parasite life stage. The *P. falciparum* blood-stage appears to contain multiple T cell determinants since different T cell clones reacted with distinct parasite antigens ranging in molecular weight from 20,000 to 200,000 kDa[56,98] and showed differential patterns of reactivity to antigens present in the heat-stable and heat-precipitated fractions of lysates of *P. falciparum*-infected erythrocytes.[98] The identification and relative importance of these various T cell antigens in the induction of a protective immune response remain to be established. All of the parasite-specific human T cell clones obtained by various groups using stimulation of T cells with antigen and IL-2 appeared to be of the CD4[+] phenotype (helper/inducer).[97-99] However, Sinigaglia et al.[94] reported that, following an initial 5-d *in vitro* stimulation of peripheral T cells with parasite antigen, the use of nonselective culture conditions (stimulating with PHA and allogeneic feeder cells) for the cloning and maintenance of antigen-sensitized T cells allowed for the isolation of cells with both CD4[+] and CD8[+] phenotypes. Interestingly, CD8[+] T cell clones were obtained only from acutely infected patients and not from individuals who had recovered from *P. falciparum* malaria.[94] Both CD4[+] and CD8[+] human T cell clones were shown to proliferate and produce IFN-γ in response to stimulation with parasite antigen.[56,97] As discussed in the next section, IFN-γ may play an important role in the immune response to malaria due in part to its ability to activate effector macrophages. CD4[+] T cell clones, as may be expected from their phenotype, displayed helper activity in the production of anti-*P. falciparum* antibodies when cultured with autologous B-enriched peripheral lymphocytes.[94] The helper activity of CD4[+] T cell clones is presumably one of the protective functions of T lymphocytes in the host. CD8[+] T cell clones tested for cytotoxic activity failed to lyse parasitized erythrocytes or autologous EBV-transformed B cells incubated with malarial antigen.[94] However, CD8[+] clones were able to lyse unrelated tumor cell targets in the presence of "bridging" anti-CD3 antibodies thus indicating that they possessed cytotoxic potential. The exact role of CD8[+] T cells in resistance to malaria remains to be determined.

Transfer experiments, which cannot be carried out in humans, can provide useful information as to the activity of selected cell populations such as T cell lines and clones.

Therefore, we took advantage of the *P. chabaudi adami* model of murine malarial infection to perform such experiments. As mentioned earlier, recovery from infection with this parasite is mediated by antibody-independent, T cell-mediated immune mechanisms. *P. chabaudi adami*-specific T cell lines were developed by Brake et al.[55] by alternate stimulation of splenic T cells from immune animals with antigen and IL-2. The T cell lines obtained were L3T4[+] and proliferated in an H-2 restricted manner when incubated with parasite antigen and antigen-presenting cells.[55] Adoptive transfer experiments showed that such T cell lines were able to protect both athymic nude mice and sublethally irradiated recipients against *P. chabaudi adami* infection, i.e., the animals developed low levels of parasitemia followed by clearance of the parasite as opposed to a fatal infection in control mice. These results unequivocally show that, in *P. chabaudi adami* infections, T cells of helper phenotype (L3T4[+]) specific for parasite antigen play a primary role in recovery from blood-stage infection independent of antibody production. Mice reconstituted with T cell lines also appeared to possess specific immunologic memory since, after a primary *P. chabaudi adami* infection, they were resistant to challenge with the homologous parasite but fully susceptible to infection with the heterologous parasite *P. yoelii* 17X.[55] These results differ from the earlier findings of Gross et al.[91] who observed nonspecific protection by adoptively transferred T cell lines. In their study, an anti-PPD as well as an anti-*P. berghei* T cell line were able to protect rats against a *P. berghei* infection.[91]

In order to further characterize the functional properties and ultimately, the parasite epitopes recognized by protective T cells, Brake et al.[96] developed T cell clones from anti-*P. chabaudi adami* T cell lines by limiting dilution. Of the clones tested (CTR2.1), one showed adoptive protective activity when transferred to nude mice. This protective T cell clone displayed the L3T4[+]Lyt2[-] phenotype and produced IFN-γ and IL-2 upon antigen stimulation.[96] As observed with T cell lines, nude mice reconstituted with CTR2.1 were resistant to reinfection with *P. chabaudi adami* but not with the heterologous parasite *P. yoelii* 17X.[96] The clonality of the transferred CTR2.1 population was demonstrated by the inability of grafted mice to develop sensitization to dinitrofluorobenzene, a potent mediator of DTH, or to mount an antibody response to the unrelated antigen KLH. Moreover, subclones derived from CTR2.1 also displayed protective activity.[96] Therefore, it appears that recognition by a T cell clone of a single protective epitope expressed on the parasite is sufficient to cause the induction of protective activity against infection. As observed with splenic T cells,[80] protective T cell lines[55] and clones[96] were unable to suppress the initial ascending parasitemia suggesting that, following the recognition of antigen, the cells do not directly inhibit the parasite but rather act to activate host effector mechanisms. This could conceivably be accomplished through the production of "messenger" lymphokines such as IFN-γ and IL-2 found to be released by the protective CTR2.1 clone.

The further characterization of lymphokines produced by protective T cell clones as well as their biological activities should lead to a better understanding of the host immune responses involved in resistance to malaria. Moreover, specific T cell clones provide a unique tool for the identification and characterization of protective T cell epitopes.

IV. MECHANISMS OF T CELL-MEDIATED IMMUNITY

A. DIRECT INTERVENTION BY T LYMPHOCYTES

While there is ample evidence for the participation of T lymphocytes in protective immunity to malaria, the actual mechanisms by which the lymphocytes are able to control the growth of the parasites and resolve infection remain to be defined. T cells could conceivably act directly by cytolysis of infected erythrocytes or hepatocytes or via the production of inhibitory or toxic factors such as IFN-γ and lymphotoxin. Studies on the effect of IFN-γ indicate that this lymphokine does not directly inhibit the growth of blood-stage parasites *in vitro*.[100] *In vivo*, the administration of recombinant IFN-γ has yielded conflicting results; Clark et al.[101] reported a protective effect against murine *P. chabaudi adami* infection while Playfair and DeSouza[102] failed to observe any protection against *P. yoelii* infection in mice receiving recombinant IFN-γ. The two groups used slightly different experimental protocols which may explain differences in the outcome. In contrast to the studies on blood-stage infections, IFN-γ has shown an undisputable inhibitory effect on the hepatic stage of the parasite (see below). In the case of lymphotoxin, definite data concerning its effect on the malaria parasite are still awaited.

The direct lysis of parasitized erythrocytes and concomitant killing of parasites appears unlikely since the cytolytic process involves recognition by T lymphocytes of parasite antigens in association with MHC class I antigens which are not expressed on human erythrocytes.[103] Indeed, attempts to demonstrate cytotoxic activity of T cells against parasitized erythrocytes have generally failed. There have been reports[104,105] that splenic T cells from mice and rats infected with *P. berghei* were able to lyse parasitized erythrocytes which, in rodents, are believed to express class I MHC antigens.[103] However, these studies did not exclude the possibility that the cytotoxicity observed may have been mediated by contaminating cell populations or by mechanisms other than cytolysis. In addition, Jayawardena et al.[63] and Brinkmann et al.[85] have shown that, unlike unfractionated T cells or Lyt1⁺ cells, Lyt2⁺ cells failed to transfer protection against blood-stage *P. yoelii* in mice thereby arguing against the involvement of cytotoxic T cells in the control of parasitemia. On the other hand, recent evidence suggests that, in the case of malarial infections initiated by sporozoite inoculation, CD8⁺ T lymphocytes (cytotoxic/suppressor phenotype) may play a major role in protection against the parasite.[106,107] Schofield et al.[106] and Weiss et al.,[107] studying infections with sporozoites from *P. berghei* and *P. yoelii*, respectively, reported that sporozoite-immunized mice which would normally resist reinfection lost their immune status when depleted of CD8⁺ T cells by injection of CD8-specific monoclonal antibodies. Similar removal of CD4⁺ T cells did not reduce immunity. Following sporozoite inoculation, the development of liver exoerythrocytic forms was prevented in intact immune animals[106] and, therefore, it was suggested that CD8⁺ T cells exerted their action at the level of the hepatocyte. The authors speculated that recognition by CD8⁺ T cells of parasite antigens in the context of MHC class I antigens expressed on the surface of hepatocytes may lead to direct cytolysis of infected liver cells or to the inhibition of parasite growth through the release of IFN-γ by the activated CD8⁺ T cells.[106,107] Consistent with this proposition, IFN-γ has been shown to inhibit the development of exoerythrocytic forms in liver cells both *in vivo*[108,109] and *in vitro*[108,110,111] although it failed to interfere with the erythrocytic stages of the parasite.[108] Therefore, in a natural infection, sensitized CD8⁺ T cells could represent a first immunological barrier following the introduction of sporozoites through a mosquito bite. The lymphocytes could act in conjunction with antisporozoite antibodies which have been shown to inhibit the invasion of liver cells by the parasite.[112,113] In accordance with this hypothesis, Schofield et al.[106] reported that naive mice receiving anti-*P. berghei* sporozoite IgG alone or immune T lymphocytes alone were only partially protected against infection and that the level of protection was significantly improved when both immune IgG and immune T lymphocytes were transferred.

B. ACTIVATION OF EFFECTOR CELLS BY T LYMPHOCYTES

While the above scenario is believed to take place at the hepatic stage of the life cycle of the parasite, T lymphocytes are also involved in protective immunity at later erythrocytic stages as demonstrated by the ability of immune T cells to promote recovery from blood-stage infection in various animal models (see Section III.D). It is believed that the release of lymphokines such as IFN-γ, macrophage chemotactic factor, IL-2, and perhaps IL-3 from T cells activated by plasmodial antigens may promote the recruitment and activation of effector cells (NK cells, polymorphonuclear leukocytes, macrophages) capable of destroying the parasite.

1. NK Cells

NK cells have the ability to lyse various tumor target cells without prior antigen sensitization and it has been proposed that they may also participate in immunity to malaria through the lysis of parasitized erythrocytes and perhaps form a first line of defense against the parasite prior to the development of an immune response.[114,115] Later during the course of infection, IL-2, and IFN-γ produced by activated T cells[116] could contribute to the enhancement of the cytolytic activity of NK cells[117,118] and the appearance of antibodies could promote the ADCC activity of NK cells.[20] The suggestion of a role for NK cells in resistance to malaria originally arose from the observation that mouse strains with high levels of splenic NK activity were resistant to plasmodial infections while mice with low levels of NK activity (A strain) were highly susceptible to infection.[115] Subsequent reports of increased NK activity during the course of human[119,120] and rodent[114,115,121,122] malarial infections lent support to the theory of NK-mediated killing of malaria parasites. However, substantial evidence argues against this possibility. For instance, the course of plasmodial infection was found to be normal in beige mice which are deficient in NK cell activity as well as in mice depleted of NK cells by treatment with ^{89}strontium or ^{17}β-estradiol.[123,124] Nude mice, in contrast, have high NK cell activity[125] but nevertheless develop unremitting parasitemias that result in death when infected with various plasmodial species.[16,80] Moreover, genetic experiments have failed to link NK cell activity and resistance to *P. chabaudi* infection in mice.[123]

2. Polymorphonuclear Leukocytes (PMNs)

While NK cells may not play a major role in the control of malarial infections, PMNs represent another population of cells whose activity does not require prior exposure to antigen and which may inhibit propagation of the parasite before the host develops a specific immune response. Phagocytosis by human PMNs of *P. falciparum*-free merozoites,[126,127] parasitized erythrocytes,[126,128,129] schizonts,[130] and gametocytes[131] has been reported and may contribute to protection. Recent studies have demonstrated that neutrophils can inhibit the asexual multiplication of *P. falciparum in vitro*.[126,130,132] Phagocytosis of parasitized erythrocytes and free parasites was observed in the cultures[126,130] but the presence of degenerating non-phagocytized parasites[126] suggested that extracellular killing of the organisms by neutrophils also contributed to the growth inhibition of *P. falciparum*. The addition of scavengers of the toxic oxygen metabolites superoxide anion and hydrogen peroxide released upon the respiratory burst had essentially no effect on the killing ability of neutrophils.[126,132] In contrast, histidine and tryptophan, both quenchers of singlet oxygen, were able to largely prevent parasite suppression by activated neutrophils suggesting that this oxygen species may be involved in parasite death.[132] In addition, it was proposed that other antimicrobial degranulation products such as lactoferrin may be involved in the destruction of *P. falciparum*.[126] This suggestion is supported by the results of Waters et al.,[133] who reported that granule proteins from a different population of PMNs, i.e., eosinophils, inhibited the multiplication of *P. falciparum* in culture and caused the appearance of abnormal parasites within erythrocytes. A monoclonal antibody against eosinophil cationic protein partially blocked the inhibitory effect indicating that, although this molecule contributed to parasite death, other degranulation products (e.g., arylsulfatase B, phospholipase D, lysophospholipase, etc.) are probably involved.[133] These data are in accordance with the view that PMNs act as a nonspecific cellular immune defense mechanism against malaria. During the course of infection, the parasiticidal activity of PMNs could be enhanced by antibodies (promotion of phagocytosis, activation) and modulated by T cells via lymphotoxin which has been reported to activate and be chemotactic for PMNs.[134]

3. Macrophages

Since the early work of Taliaferro and Mulligan,[135] it has been known that macrophages are prominent in malarial infections. In response to circulating parasitized erythrocytes, the malarious host displays a dramatic increase in blood monocytes and the accumulation of macrophages in the liver and the spleen.[136] Circulating monocytes and macrophages from these organs appear to be activated as demonstrated by marked changes in surface phenotype, and secretory and phagocytic abilities.[136,137] The phagocytic activity of macrophages is believed to contribute to the clearance of parasitized red blood cells by the liver and the spleen during a malaria infection[23,135] as suggested by the presence of free parasites, parasitized erythrocytes, uninfected erythrocytes, malaria pigment, and erythrocytic debris inside the macrophages present in these organs.[135,138] Nevertheless, animal models have shown that enhanced phagocytosis, although an important component of immunity in malaria, does not appear to play a decisive role in recovery from infection.[139-141] It has been suggested that a major protective function of macrophages/monocytes consists of their ability to release, upon activation, factors that are cytotoxic for blood-stage parasites.[142,143] Macrophage activation can be induced by lymphokines such as IFN-γ released by T lymphocytes following recognition of specific plasmodial antigens or stimulation by mitogenic components of the parasite.[138,144-147] T cells have been reported to accumulate in the same locales as macrophages during infection, i.e., spleen and liver, where these two cell types would therefore be brought into contact.[48,59] In the presence of antibodies, macrophage activation can also result from phagocytosis of immune complexes and opsonized parasites.[145,148-150] In addition, antibodies may reinforce the effects of macrophage-mediated immunity by targeting parasitized erythrocytes or free parasites onto effector macrophages which express surface Fc and C3b receptors.

There is a large body of evidence indicating that activated macrophages release factors toxic to the malaria parasite. The mediators implicated so far include reactive oxygen species (ROS) and the lipid peroxidation products generated by these molecules, factors present in tumor necrosis serum, and possibly, enzymes such as polyamine oxidase. Such molecules released by effector macrophages have been reported to cause intraerythrocytic death of the parasite thus giving rise to "crisis forms" consisting of degenerate parasites inside apparently intact erythrocytes.[100,138,145,146,151] Crisis forms were first described by Taliaferro and Taliaferro[152] in 1944, who observed them in the blood of Cebus monkeys during a rapid decrease in parasitemia. Subsequent reports indicated that the appearance of crisis forms within circulating erythrocytes correlated with recovery from murine and simian malarial infections.[23,153] Therefore, the activity and mode of action of the growth-inhibitory factors produced by macrophages have been the subject of intense investigation. Results from such studies are briefly reviewed in the next sections.

a. Oxidative Killing

Allison and Eugui[20,142] suggested that effector cells activated by T lymphocytes released reactive oxygen intermediates which exerted oxidant stress on parasitized erythrocytes and caused intraerythrocytic death of the parasite. It was argued that the sensitivity of malaria parasites to oxidant stress was illustrated in nature by the reduced ability of *P. falciparum* to survive in human erythrocytes with increased susceptibility to oxidative damage due to abnormal hemoglobins or deficiencies in glucose-6-phosphate dehydrogenase or vitamin E.[154-156] Their interpretation is supported by evidence obtained both *in vivo* and *in vitro*. It has been shown that, during the course of malarial infection in rodents[157-159] and humans,[160] macrophages are indeed activated to release oxygen radicals and that there may be a correlation between the capacity of macrophages to produce oxygen radicals and the ability to recover from infection.[157,159] This contention is supported by experiments showing that ROS released during the oxidative burst of activated macrophages[145] or generated by enzyme-substrate systems[145,161-163] are able to kill human[161,163] and murine[145,162] intraerythrocytic parasites. Various highly reactive and potentially toxic oxygen species are produced during the oxidative burst, including superoxide anion (O_2^-), hydrogen peroxide (H_2O_2), singlet oxygen, and hydroxyl radical. However, the major effector molecule against the parasite appeared to be H_2O_2. For instance, the killing of intraerythrocytic *P. yoelii* by products of activated macrophages was inhibited by catalase, a scavenger of H_2O_2, while the presence of superoxide dismutase, histidine, or mannitol which are scavengers of O_2^-, singlet oxygen, and hydroxyl radical, respectively, was unable to prevent killing of the parasite.[145] Similarly, in enzyme-substrate systems, blood-stage *P. yoelii*[145,162] and *P. falciparum*[161] were highly susceptible to H_2O_2, the only product generated by the interaction of glucose-glucose oxidase and the killing was reversed by the addition of exogenous catalase. Products of the xanthine-xanthine oxidase system which, in addition to H_2O_2, include O_2^-, hydroxyl radicals, and singlet oxygen, were also lethal to the parasites,[145,161,162] but only catalase was able to inhibit the killing.[145,162] Scavengers of other oxygen radicals such as superoxide dismutase, mannitol, and histidine had no effect.[145,161,162] Finally, Dockrell and Playfair[163] showed that concentrations of H_2O_2 as low as $10^{-5} M$ were directly toxic to blood-stage *P. yoelii* and *P. berghei in vitro*. *In vivo* studies in which animals were injected with H_2O_2[143,163] or various agents known to generate free oxygen radicals such as phenylhydrazine,[143,164] alloxan,[143] *t*-butyl hydroperoxide,[142,165,166] or divicine[167] confirmed the importance of ROS in causing intraerythrocytic parasite death and control of parasitemia. Injection of these compounds into parasitized animals resulted in a rapid reduction in parasitemia accompanied by hemolysis and the appearance of degenerate parasites (crisis forms) within circulating erythrocytes.[142,143,163,165-167] The injection of desferrioxamine, a specific iron chelator which inhibits the iron-catalyzed formation of hydroxyl radicals from O_2^- and H_2O_2,[168] prevented the inhibitory effect of alloxan,[143] *t*-butyl hydroperoxide,[142,165,166] and divicine[167] thereby substantiating the role of ROS in the killing of intraerythrocytic plasmodia by these agents. Furthermore, Clark et al.[101] have recently shown that during the course of an undisturbed *P. chabaudi* infection, mice fed a diet containing butylated hydroxyanisole, a scavenger of free radicals, suffered more severe infections with higher parasitemias and delayed appearance of crisis forms. This observation brings further support to the hypothesis of oxygen-mediated killing of malaria parasites.

While the ability of products of the oxidative burst to affect parasite growth appears well established, the actual mechanisms by which these intermediates effect killing of plasmodia remain to be defined. Intraerythrocytic parasites appear themselves to exert oxidant stress on the host erythrocyte as evidenced by the increased lipid peroxidation of parasitized erythrocyte membranes[166,169] and their increased sensitivity to lysis following exposure to oxidative agents compared to normal erythrocytes.[143,166,167] Therefore, the decreased parasitemia observed *in vivo* following the injection of radical-generating compounds[142,143,165-167] is likely to be, at least in part, due to direct lysis of parasitized erythrocytes already under oxidative stress. The frequent observation of degenerating parasites inside intact circulating erythrocytes following exposure to oxidative systems *in vivo*[142,143,163,165-167] and *in vitro*[145,161-163] implies that parasite death is not solely a passive consequence of hemolysis. It has been shown that oxygen radicals can traverse erythrocyte membranes[170] and it is possible that, once inside, they can damage the parasite directly. It has also been suggested that the alteration in the permeability of the host erythrocyte and consequent loss of K^+ ions caused by oxidative agents may contribute to inhibit parasite growth.[154] Lipid peroxidation by ROS may also contribute to parasite death by reducing the deformability of the host erythrocyte and thus its *in vivo* survival.[171] More importantly perhaps, radical-induced lipid peroxidation can result in the generation of aldehydes toxic to the human malaria parasite.[172,173] Aldehydes have longer half-lives than oxygen radicals and could affect parasites not only locally, but also at more distant sites.[172-174]

A general picture emerges from the evidence summarized in this section; upon recognition of specific malarial antigens, T lymphocytes release lymphokines which recruit[65] and activate macrophages in the liver, spleen, and peripheral blood. Macrophage activation can also result from phagocytosis which can be enhanced by the presence of antiplasmodial antibodies. The small blood vessels of the liver and the spleen present an optimal environment for the trapping and destruction of parasites as they are brought in close contact with populations of highly activated macrophages releasing toxic oxygen species. Here again, antibodies can synergize with CMI by opsonizing parasitized erythrocytes and maximizing contact between targets and effector cells.[145,157] The release of oxygen radicals and toxic aldehydes could also contribute to the tissue injury observed in malaria.[143,172,174]

b. Tumor Necrosis Factor, Crisis Form Factor

Despite the demonstrated importance of the oxidative killing of plasmodial species, other factors appear to be involved in parasite destruction as leukocytes obtained from patients with chronic granulomatous disease are able to kill blood-stage *P. falciparum in vitro* despite their inability to generate a respiratory burst.[146,175] One nonoxidative factor generated by macrophages which has received a great deal of attention is tumor necrosis factor (TNF). As mentioned earlier, (Section III.A), macrophage-activating agents such as *C. parvum* and BCG can induce nonspecific protection of mice against various species of Plasmodium and Babesia.[35,36] Crisis forms were observed in the circulation of the protected animals and it was proposed that these agents acted by inducing the release of soluble parasiticidal factors.[35,36] Indeed, when animals given *C. parvum* or BCG are subsequently injected with endotoxin, a tumoricidal factor is released by activated macrophages that can be found in the serum (tumor necrosis serum or TNS[176]). Such TNS has been shown to suppress parasite growth both *in vivo* and *in vitro* and to induce the development of crisis forms.[100,151,161,177-179] Clark[180] proposed that a similar sequence of events was taking place during the course of malarial infection; Wood and Clark[181] showed that infection with hemoprotozoa resulted in the activation of macrophages and their "priming" for the release of monokines, much in the same fashion as an injection of BCG, except that in a malarial infection macrophage activation was presumably due to the release of IFN-γ by specifically activated T cells. It was suggested that subsequently, the production of tumoricidal/parasiticidal factors by macrophages could be triggered either by endotoxin from bacteria of the indigenous flora of the host or by an endotoxin-like substance present in the parasite itself.[180,182] The macrophage factor(s) released would then mediate intraerythrocytic death of the parasite and possibly some of the pathological changes which can accompany malaria and resemble those seen in endotoxic shock (e.g., hypoglycemia, adrenal damage, consumptive coagulopathy, etc.)[180,182] This hypothesis was supported by the observation that in mice, the injection of endotoxin during the course of infection with *P. vinckei petteri*[183] or *P. yoelii*[178] as an external "trigger" for the release of TNS factor(s) by macrophages resulted in growth inhibition of the parasite in LPS responder mice[178] but not in LPS nonresponder mice.[183,184] Reports of endotoxin-like activity in the serum of humans infected with *P. falciparum* and mice infected with *P. berghei*,[185] as well as raised levels of TNF in malarious patients,[186] suggest that the "endotoxin trigger" for the production of TNS factors may be present in an undisturbed *in vivo* situation. In addition, it has recently been shown that IL-2 and IFN-γ, which are both produced by T cells in response to malarial antigens,[116] can synergize to induce the production of TNF.[187]

The parasiticidal activity of TNS was initially attributed to TNF[151,161,179] but it has since become evident that recombinant TNF/cachectin does not directly affect the growth of malaria parasites *in vitro*[188,189] indicating that TNS contains other factor(s) active against the parasite. By contrast, in *in vivo* studies, repeated injections of recombinant TNF[190] or its continuous delivery from an i.p. osmotic pump[101] did succeed in reducing parasitemia suggesting that the effect of TNF on blood-stage parasites is indirect, most likely through the induction of a host effector mechanism.[101,190] In accordance with this, there are reports that TNF can activate eosinophils[191,192] and neutrophils[193] which, as mentioned in Section IV.B.2, are able to act as effector cells against the parasites.

The identity of the factor(s) present in TNS that are directly cytotoxic to parasites remains to be determined. It could conceivably consist of certain macrophage enzymes such as polyamine oxidase which, in the presence of substrate (spermine or spermidine), generates H_2O_2 and aldehydes[194] and which has been reported to inhibit the growth of *P. falciparum in vitro*.[195] The parasiticidal component(s) of TNS could also be related or identical with the crisis form factor (CFF) described by Jensen et al.[196,197] that is present in the serum of Sudanese residents of malaria-endemic areas. Sera from these individuals inhibited the intraerythrocytic development of *P. falciparum* and gave rise to the appearance of crisis forms.[188,186] The CFF present in those sera was shown to be different from antibodies or TNF[188,196,197] and has yet to be characterized.

V. ROLE OF THE SPLEEN IN CELL-MEDIATED IMMUNITY TO MALARIA

Hepatosplenomegaly has long been considered a hallmark of malaria and it is recognized that splenectomy usually enhances the severity of this disease.[98] In fact, the significance of the spleen in resistance to malaria is evidenced by the demonstration that the characteristically narrow host range of plasmodia was extended by the splenectomy of unnatural hosts. While the exact nature of the role played by the spleen in resistance to malaria remains to be determined, in all likelihood it is very complex and involves an array of capabilities normally associated with spleen function.[199] These include the filtration of blood through the spleen with the recruitment, compartmentalization, and storage of various blood cells; the removal and destruction of aged or damaged blood cells including parasitized erythrocytes; the providing of a hemopoietic microenvironment where various blood cells undergo final differentiation or, in certain species, an anatomic location for hematopoiesis including erythropoiesis; and the creation of an environment where cells of the immune system interact with antigens and each other to produce cells and molecules playing a role in immunity.

During malaria the spleen enlarges dramatically in euthymic hosts but less so in athymic animals suggesting that the phenomenon of splenomegaly in response to malaria is dependent upon T cell activation.[200] Moreover, the activation and recruitment of monocytes from bone marrow to spleen appears to be T cell-dependent as indicated above, and it is likely that increased blood levels of IFN-γ seen during malaria[66,67] result from the recruitment to and activation of T cells within the spleen and liver of the infected host. Large numbers of activated lymphoid cells and macrophages have been reported to accumulate in both organs during malaria.[136,201,202] In the case of *P. yoelii* infection, it appears that CMI in the liver is associated with recovery.[25] In recent studies, Taverne et al.[203] reported that macrophages in the livers of mice infected with virulent *P. yoelii* produce an enhanced oxidative burst and display increased cytotoxicity for tumor cells. These characteristics were especially marked in immunized mice which recovered from infection. We have already commented on the vital role played by CMI in resistance to *P. chabaudi adami* malaria. When B cell-deficient mice were splenectomized prior to infection with this parasite or following the termination of acute disease, they developed unremitting malaria with high parasitemias and eventually died.[204] These data suggest that the expression of CMI to *P. chabaudi adami* malaria in this model requires the presence of a spleen. In contrast, immunologically intact animals which had been splenectomized prior to infection suffered malaria of prolonged duration and eventually cleared parasites from their blood more than 2 months after non-splenectomized control mice had resolved their malaria. Mice which were splenectomized and reconstituted with a monodispersed spleen cell equivalent prior to infection developed malaria similar to that of nonreconstituted splenectomized mice. These data indicate that the expression of CMI to acute *P. chabaudi adami* malaria is dependent upon the presence of an architecturally intact spleen. The fact that immunologically intact animals which were splenectomized prior to infection resolved their malaria, combined with the observation that immunologically intact mice which had been immunized by active infection and then splenectomized were resistant to challenge infection clearly indicate that additional mechanisms of immunity are expressed in immunologically intact mice.

Many events take place in the spleen during malaria including changes in histological, hematopoietic, immunological, and physiological characteristics.[136,205,206] Changes occur in blood filtration as well as blood circulation within the spleen and in part depend upon the host-parasite model being studied, the immunologic status of the host and even the time period during infection chosen for study.[207,208] It is no small wonder that we do not know exactly how the spleen functions in resistance to malaria. As suggested by others,[8,9,20] parasite destruction probably occurs within the spleen and the liver via CMI mechanisms which involve the stimulation of antigen-reactive T cells by plasmodial determinants processed by antigen-presenting cells. These lymphoid cells replicate, differentiate, and secrete effector molecules which serve to recruit and activate other cells including lymphocytes, monocytes/ macrophages, polymorphs, hematopoietic stem cells, NK cells, etc. Certain of these cells produce molecules such as those described above and possibly others which remain to be discovered that are toxic for the parasites. Other cells may express adhesion molecules which could facilitate the adherence of parasites and parasitized erythrocytes to membrane surfaces where they would be exposed to toxic molecules or phagocytized by activated cells within these organs. As mentioned earlier, antibodies may serve to enhance these activities by increasing contact between effector and target cells in addition to other antiparasitic activities. We have much to learn about the cells and molecules involved in these events as well as the parasite antigens which initiate these interactions and serve as targets of CMI.

VI. SIGNIFICANCE OF T CELL EPITOPES IN THE DESIGN OF PEPTIDE VACCINES

In light of the demonstrated involvement of T lymphocytes in resistance to malaria, it becomes obvious that optimal levels of protective activity by immunization with recombinant or synthetic peptides will require the inclusion of not only B cell epitopes, but also T cell epitopes capable of inducing T cell helper and effector activities. This has already been confirmed in recent studies on the protective activity of sporozoite vaccines consisting of subunit peptides of the CS protein. As already mentioned, Egan et al.[89] showed that, in the mouse, subunit vaccines elicited humoral reponses that were equivalent to or greater than those elicited by irradiated whole sporozoites. Nevertheless, the protection against sporozoite challenge induced by subunit vaccines was far less than that achieved by attenuated sporozoites. Transfer studies by the same group[89] as well as others[88] showed that animals immunized with whole sporozoites developed, in addition to antibodies, CMI that could be transferred by T lymphocytes to naive animals. These results are in accordance with the earlier findings of Chen et al.,[31] who demonstrated that protective immunity could be induced by immunization with irradiated sporozoites in B cell-deficient (u-suppressed) mice but not in T cell-deficient animals. These animal studies indicated that effective immunizing agents should contain epitopes recognized not only by T helper cells but also by T cells capable of inducing antibody-independent cell-mediated mechanisms of immunity.

The requirement for epitopes capable of inducing T cell effector in addition to T helper activity in potential vaccines may be in part responsible for the limited effectiveness of the sporozoite subunit vaccines tested so far in humans.[6,7] A 12-amino acid synthetic peptide

(NANP)₃ comprising the immunodominant B cell epitope of the *P. falciparum* CS protein conjugated to tetanus toxoid (tet)[6] as well as a recombinant vaccine containing sequential repeats of the same dominant CS protein B cell epitope followed by the first 32 amino acids encoded by a tetracycline resistance gene read out of frame (R32tet32)[7] were used by different groups to immunize volunteers against natural infection with sporozoites. In both cases, the majority of volunteers developed significant levels of anti-CS protein antibodies but repeated injections failed to significantly increase the titer.[6,7] When seropositive individuals were challenged with sporozoites through the bite of *Anopheles* mosquitoes, only 1/3 who received the (NANP)₃tet vaccine[6] and 1/6 immunized with the R32tet32 vaccine[7] were fully protected against the parasite. The other volunteers experienced longer prepatent periods but nevertheless developed parasitemia. Ballou et al.[7] suggested that the inclusion of additional helper epitopes in such vaccines may render them more immunogenic and generate higher, more effective levels of protective antibodies as well as improve the memory component of the immune response. The inclusion in subunit vaccines of epitopes capable of inducing antibody-independent T cell-mediated immunity (e.g., activation of CD8⁺ T cells against sporozoites) would also be likely to increase their effectiveness.

Carrier proteins such as tetanus toxoid in the (NANP)₃tet vaccine or fusion products such as the tet32 tail in R32tet32 can provide epitopes for T effector cells as well as helper T cells that can improve antibody production against the parasite antigen portion of the molecule. However, the development of sensitized T cells specific for the foreign portion of a fusion product or the carrier part of a subunit vaccine would fail to give rise to boosting of the immune response upon subsequent natural infection since foreign epitopes present in the vaccine are not expressed on the parasite. The importance of this consideration is well illustrated by the studies of Good et al.,[209] who reported that in the mouse, only two out of nine strains of mice whose T cells responded to R32tet32 and provided helper activity actually recognized epitopes present in the malaria-encoded sequence (R32). The remainder responded to T cell epitopes present in the tet32 tail. Therefore, to insure natural boosting, both B and T cell epitopes covalently linked in a peptide vaccine should be derived from the same natural parasite protein.[209]

Several investigators[209-211] have also underlined the problem of MHC restriction in the design of subunit vaccines. They observed that only mice expressing the histocompatibility I-Aᵇ class II molecule were able to mount an antibody response[209,210] or give a T cell proliferative response[209,211] to NANP repeat peptides from the CS protein of sporozoites indicating that the ability of a given epitope to activate T cells in different MHC settings may be rather limited. Therefore, to be fully effective, a malarial vaccine should contain several T cell epitopes, at least one of which would be recognized by any given individual receiving the vaccine.

Another problem in the design of subunit vaccines regards parasite polymorphism. Good et al.[212] recently defined three immunodominant sites located outside the constant repeat region of the CS protein of *P. falciparum* that were able to stimulate proliferation of peripheral blood lymphocytes from adults living in a *P. falciparum* malaria-endemic region of West Africa. Unfortunately, the immunodominant T cell epitopes coincided with the polymorphic regions of the protein making inclusion of these regions in a vaccine of limited value.[212] The authors suggested that parasite mutation and selection in response to immune pressure from T cells were responsible for this phenomenon.[212] The inclusion of multiple variant sequences in future malarial vaccines may be necessary to answer the problem of polymorphism within major T cell domains.

To summarize, an efficacious malaria vaccine should ideally contain

1.   Covalently linked B and T helper cell epitopes represented on the same natural parasite protein (to insure natural boosting)

2. Epitopes recognized by T cells involved in antibody-independent effector mechanisms of immunity
3. T cell epitope(s) recognized by individuals of different MHC types
4. Variants of a given T cell determinant expressed by different strains of the parasite to which the vaccinee may be exposed
5. The vaccine should lack suppressor T cell epitopes

The identification of T cell epitopes to be incorporated in a potential peptide vaccine can be carried out by various methods. Antibodies have already been used to detect parasite molecules which may contain T cell as well as B cell epitopes (e.g., repeat region of the CS protein).[213] Assays of T cell proliferation[209,211,212] or T helper activity[109,210] as well as induction of DTH[214] upon administration of defined peptide antigens can also be used to identify T cell epitopes. Computer models have also been successfully used to predict T cell sites on protein molecules.[215] Finally, parasite-specific T cell clones with well-characterized biological activities may provide the ultimate tool for the identification of potentially protective T cell epitopes.

ACKNOWLEDGMENTS

We wish to thank Donna M. Russo, Lisa A. Cavacini, and James M. Burns, Jr., for their editorial assistance. Financial support was provided by grant no. AI12170 from the National Institutes of Health. Johanne Melancon-Kaplan is the recipient of a fellowship from the Medical Research Council of Canada.

REFERENCES

1. **Neva, F. A.,** Looking back for a view of the future: observations of immunity to induce malaria, *Am. J. Trop. Med. Hyg.,* 16, 211, 1977.
2. **Cohen, S. and Lambert, P. H.,** Malaria, in *Immunology of Parasitic Infections,* 2nd ed., Cohen, S. and Warren, D., Eds., Blackwell, London, 1982, 422.
3. **Cohen, S.,** Immunity to malaria, *Proc. R. Soc. Ser. B.,* 203, 323, 1979.
4. **Brown, G. V. and Nossal, G. J. V.,** Malaria-yesterday, today, and tomorrow, *Perspect. Biol. Med.,* 30, 65, 1986.
5. **Miller, L. H., Howard, R. J., Carter, R., Good, M. F., Nussenzweig, V., and Nussenzweig, R.,** Research towards malaria vaccines, *Science,* 234, 1349, 1986.
6. **Herrington, D. A., Clyde, D. F., Losonsky, G., Cortesia, M., Murphy, J. R., Davis, J., Baqar, S., Felix, A. M., Heimer, E. P., Gillessen, D., Nardin, E., Nussenzweig, R. S., Nussenzweig, V., Hollingdale, M. R., and Levine, M. M.,** Safety and immunogenicity in man of a synthetic peptide malaria vaccine against *Plasmodium falciparum* sporozoites, *Nature,* 328, 257, 1987.
7. **Ballou, W. R., Sherwood, J. A., Neva, F. A., Gordon, D. M., Wirtz, R. A., Wasserman, G. F., Diggs, C. L., Hoffman, S. L., Hollingdale, M. R., Hockmeyer, W. T., Schneider, I., Young, J. F., Reeve, P., and Chulay, J. D.,** Safety and efficacy of a recombinant DNA *Plasmodium falciparum* sporozoite vaccine, *Lancet,* 8545, 1277, 1987.
8. **Clark, I. A.,** Cell-mediated immunity in protection and pathology of malaria, *Parasit. Today,* 3, 300, 1987.
9. **Playfair, J. H. L.,** Immunity to malaria, *Br. Med. Bull.,* 38, 153, 1982.
10. **Deans, J. A. and Cohen, S.,** Immunology of malaria, *Annu. Rev. Microbiol.,* 37, 25, 1983.
11. **Brown, K. N., Berzins, K., Jarra, W., and Schetters, T.,** Immune responses to erythrocytic malaria, *Clin. Immunol. Allerg.,* 6, 227, 1986.
12. **Brown, I. N., Allison, A. C., and Taylor, R. B.,** *Plasmodium berghei* infections in thymectomized rats, *Nature,* 219, 292, 1968.
13. **Carter, R. L., Leuchars, E., Davies, I., and Stechschulte, D. J.,** *Plasmodium berghei* infection in thymectomized rats, *Proc. Soc. Exp. Biol. Med.,* 131, 748, 1969.

14. **Jayawardena, A. N., Targett, G. A. T., Carter, R. L., Leuchars, E., and Davies, A. J. S.,** The immunological response of CBA mice to *P. yoelii*. I. General characteristics of the effects of T-cell deprivation and reconstitution with thymus grafts, *Immunology*, 32, 849, 1977.

15. **Spira, D. T., Silverman, P. H., and Gaines, C.,** Anti-thymocyte serum effects on *Plasmodium berghei* infection in rats, *Immunology*, 19, 759, 1970.

16. **Clark, I. A. and Allison, A. C.,** *Babesia microti* and *Plasmodium berghei yoelii* infections in nude mice, *Nature*, 252, 328, 1974.

17. **Roberts, D. W., Rank, R. G., Weidanz, W. P., and Finerty, J. F.,** Prevention of recrudescent malaria in nude mice by thymic grafting or by treatment with hyperimmune serum, *Infect. Immun.*, 16, 821, 1977.

18. **Brown, K. N.,** Protective immunity to malaria provides a model for the survival of cells in an immunologically hostile environment, *Nature*, 230, 163, 1971.

19. **Allison, A. C. and Clark, I. A.,** Specific and non-specific immunity to haemoprotozoa, *Am. J. Trop. Med. Hyg.*, 26, 216, 1977.

20. **Allison, A. C. and Eugui, E. M.,** The role of cell-mediated immune responses in resistance to malaria, with special reference to oxidant stress, *Annu. Rev. Immunol.*, 1, 361, 1983.

21. **Jayawardena, A. N., Targett, G. A. T., Leuchars, E., and Davies, A. J. S.,** The immunological response of CBA mice to *P. yoelii*. II. The passive transfer of immunity with serum and cells, *Immunology*, 34, 157, 1978.

22. **Cohen, S., McGregor, I. A., and Carrington, S.,** Gamma-globulin and acquired immunity to human malaria, *Nature*, 192, 733, 1961.

23. **Quinn, T. C. and Wyler, D. J.,** Intravascular clearance of parasitized erythrocytes in rodent malaria, *J. Clin. Invest.*, 63, 1187, 1979.

24. **Brown, I. N. and Phillips, R. S.,** Immunity to *Plasmodium berghei* in rats: passive serum transfer and role of the spleen, *Infect. Immun.*, 10, 1213, 1974.

25. **Playfair, J. H. L., DeSouza, J. B., Dockrell, H. M., Agomo, P. U., and Taverne, J.,** Cell-mediated immunity in the liver of mice vaccinated against malaria, *Nature*, 282, 731, 1979.

26. **Freeman, R. R. and Holder, A. A.,** Characteristics of the protective response of BALB/c mice immunized with a purified *Plasmodium yoelii* schizont antigen, *Clin. Exp. Immunol*, 54, 609, 1983.

27. **Grun, J. L. and Weidanz, W. P.,** Immunity to *Plasmodium chabaudi adami* in the B-cell-deficient mouse, *Nature*, 290, 143, 1981.

28. **Grun, J. L. and Weidanz, W. P.,** Antibody-independent immunity to reinfection malaria in B-cell-deficient mice, *Infect. Immun.*, 41, 1197, 1983.

29. **Ferris, D. H., Beamer, P. D., and Stutz, D. R.,** Observations on the response of dysgammaglobulinemic chickens to malaria infection, *Avian Dis.*, 17, 12, 1973.

30. **Rank, R. G. and Weidanz, W. P.,** Nonsterilizing immunity in avian malaria: an antibody-independent phenomenon, *Proc. Soc. Exp. Biol. Med.*, 151, 257, 1976.

31. **Chen, D. H., Tigelaar, R. E., and Weinbaum, F. I.,** Immunity to sporozoite-induced malaria infection in mice. I. The effect of immunization of T and B cell-deficient mice, *J. Immunol.*, 118, 1322, 1977.

32. **Strickland, G. T. and Hunter, K. W.,** The use of immunopotentiators in malaria, *Int. J. Nucl. Med. Biol.*, 7, 133, 1980.

33. **Jayawardena, A. N.,** Immune responses in malaria, in *Parasitic Diseases*, Vol. 1, Mansfield, J. M., Ed., Marcel Dekker, New York, 1981, 85.

34. **Nussenzweig, R. S.,** Increased nonspecific resistance to malaria produced by administration of killed *Corynebacterium parvum*, *Exp. Parasitol.*, 21, 224, 1967.

35. **Clark, I. A., Cox, F. E. G., and Allison, A. C.,** Protection of mice against *Babesia* spp. and *Plasmodium* spp. with killed *Corynebacterium parvum*, *Parasitology*, 74, 9, 1977.

36. **Clark, I. A., Allison, A. C., and Cox, F. E.,** Protection of mice against *Babesia* and *Plasmodium* with BCG, *Nature*, 259, 309, 1976.

37. **Allison, A. C., Christensen, J., Clark, I. A., Elcord, B. C., and Eugui, E. M.,** The role of the spleen in protection against murine Babesia infections, in *The Role of the Spleen in the Immunology of Parasitic Diseases*, Tropical Diseases Research Series, No. 1, Torrigiani, G., Ed., Schwabe & Co. AG, Basel, 1979, 151.

38. **Cox, F. E. G.,** Protective immunity between malaria parasites and piroplasms in mice, *Bull. W.H.O.*, 43, 325, 1970.

39. **Cox, F. E. G. and Turner, S. A.,** Antigenic relationships between the malaria parasites and piroplasms of mice as determined by the fluorescent-antibody technique, *Bull. W.H.O.*, 43, 337, 1970.

40. **Weidanz, W. P. and Grun, J. L.,** Antibody-independent mechanisms in the development of acquired immunity to malaria, in *Host Defenses to Intracellular Pathogens*, Vol. 162, Eisenstein, T. K., Actor, P., and Friedman, H., Eds., Plenum Press, New York, 1983, 409.

41. **Longenecker, B. M., Breitenbach, R. P., and Farmer, J. N.,** The role of the bursa of fabricius, spleen and thymus in the control of a *Plasmodium lophurae* infection in the chicken, *J. Immunol.*, 97, 594, 1966.

42. **Weinbaum, F. I., Evans, C. B., and Tigelaar, R. E.,** Immunity to *Plasmodium berghei yoelii* in mice. I. The course of infection in T cell and B cell deficient mice, *J. Immunol.,* 117, 1999, 1976.

43. **Roberts, D. W. and Weidanz, W. P.,** T-cell immunity to malaria in the B-cell-deficient mouse, *Am. J. Trop. Med. Hyg.,* 28, 1, 1979.

44. **Hunter, K. W., Finkelman, F. D., Strickland, G. T., Sayles, P. C., and Scher, I.,** Defective resistance to *Plasmodium yoelii* in CBA/N mice.I, *J. Immunol.,* 123, 133, 1979.

45. **Jayawardena, A. N., Janeway, C. A., Jr., and Kemp, J. D.,** Experimental malaria in the CBA/N mouse, *J. Immunol.,* 123, 2352, 1979.

46. **Grun, J. L. and Weidanz, W. P.,** unpublished observation, 1977.

47. **Cavacini, L. A., Parke, L. M., and Weidanz, W. P.,** unpublished observation, 1988.

48. **Jayawardena, A. N., Targett, G. A. T., Leuchars, E., Carter, R. L., Doenhoff, M. J., and Davies, A. J. S.,** T-cell activation in murine malaria, *Nature,* 258, 149, 1975.

49. **Freeman, R. R. and Parrish, C. R.,** Spleen cell changes during fatal and self-limiting malarial infections of mice, *Immunology,* 35, 479, 1978.

50. **Weinbaum, F. I., Evans, C. B., and Tigelaar, R. E.,** An *in vitro* assay for T-cell immunity to malaria in mice, *J. Immunol.,* 116, 1280, 1976.

51. **Troye-Blomberg, M., Perlmann, H., Patarroyo, M. E., and Perlmann, P.,** Regulation of the immune response in *Plasmodium falciparum* malaria. II. Antigen-specific proliferative responses *in vitro, Clin. Exp. Immunol.,* 53, 345, 1983.

52. **Theander, T. G., Bygberg, I., Jepsen, S., Svenson, M., Kharazmi, A., Larsen, P. B., and Bendtzen, K.,** Proliferation induced by *Plasmodium falciparum* antigen and interleukin-2 production by lymphocytes isolated from malaria immune individuals, *Infect. Immun.,* 53, 221, 1986.

53. **Ho, M., Webster, H., Looareesuwan, S., Supanaranond, W., Phillips, R., Chanthavanich, P., and Warrell, D.,** Antigen-specific immunosuppression in human malaria due to *Plasmodium falciparum, J. Infect. Dis.,* 153, 763, 1986.

54. **Theander, T. G., Bygberg, I. C., Andersen, B. J., Jepsen, S., Kharazmi, A., and Odum, N.,** Suppression of parasite specific response in *Plasmodium falciparum* malaria. A longitudinal study of blood mononuclear cell proliferation and subset composition, *Scand. J. Immunol.,* 24, 73, 1986.

55. **Brake, D. A., Weidanz, W. P., and Long, C. A.,** Antigen-specific, Interleukin 2-propagated T lymphocytes confer resistance to *Plasmodium chabaudi adami, J. Immunol.,* 137, 347, 1986.

56. **Sinigaglia, F. and Pink, J. R. L.,** Human T lymphocyte clones specific for malaria (*Plasmodium falciparum*) antigens, *EMBO J.,* 4, 3819, 1985.

57. **Finerty, J. F. and Krehl, E. P.,** Cyclophosphamide pretreatment and protection against malaria, *Infect. Immun.,* 14, 1103, 1976.

58. **McDonald, V. and Sherman, I. W.,** Lack of correlation between delayed-type hypersensitivity and host resistance to *Plasmodium chabaudi* infection, *Clin. Exp. Immunol.,* 42, 421, 1980.

59. **Freeman, R. R.,** T-cell function during fatal and self-limiting malarial infections of mice, *Cell. Immunol.,* 41, 373, 1978.

60. **Cottrell, B. J., Playfair, J. H. L., and DeSouza, J. B.,** Cell-mediated immunity in mice vaccinated against malaria, *Clin. Exp. Immunol.,* 34, 147, 1978.

61. **Playfair, J. H. L., DeSouza, J. B., Freeman, R. R., and Holder, A. A.,** Vaccination with a purified blood-stage malaria antigen in mice: correlation of protection with T-cell mediated immunity, *Clin. Exp. Immunol.* 62, 19, 1985.

62. **Playfair, J. H. L., DeSouza, J. B., and Cottrell, B. J.,** Protection of mice against malaria by a killed vaccine: differences in effectiveness against *P. yoelii* and *P. berghei, Immunology,* 33, 507, 1977.

63. **Jayawardena, A. N., Murphy, D. B., Janeway, C. A., and Gershon, R. K.,** T cell-mediated immunity in malaria. I. The Ly phenotype of T cells mediating resistance to *Plasmodium yoelii, J. Immunol.,* 129, 377, 1982.

64. **Russo, D. M. and Weidanz, W. P.,** *Cell. Immunol.,* Activation of antigen-specific suppressor T cells by intravenous injection of soluble blood-stage malarial antigen, *Cell. Immunol.,* 115, 437, 1988.

65. **Wyler, D. J. and Gallin, J. I.,** Spleen-derived mononuclear cell chemotactic factor in malaria infections: a possible mechanism for splenic macrophage accumulation, *J. Immunol.,* 118, 478, 1977.

66. **Eugui, E. M. and Allison, A. C.,** Natural cell-mediated immunity and interferon in malaria and babesia infections, in *NK Cells and Other Natural Effector Cells,* Herberman, R. B., Ed., Academic Press, New York, 1982, 1491.

67. **Rhodes-Feuillette, A., Bellosguardo, M., Druilhe, P., and Ballet, J. J.,** The interferon compartment of the immune response in human malaria. II. Presence of serum-interferon gamma following the acute attack, *J. Interferon Res.,* 5, 169, 1985.

68. **Troye-Blomberg, M., Andersson, G., Stoczkowska, M., Shabo, R., Romero, P., Patarroyo, E., Wigzell, H., and Perlmann, P.,** Production of IL-2 and IFN-γ by T cells from malaria patients in response to *Plasmodium falciparum* or erythrocyte antigens *in vitro, J. Immunol.,* 135, 3498, 1985.

69. **Lelchuk, R., Rose, G., and Playfair, J. H. L.,** Changes in the capacity of macrophages and T cells to produce interleukins during murine malaria infection, *Cell. Immunol.,* 84, 253, 1984.

70. **Lelchuk, R. and Playfair, J. H. L.**, Serum IL-2 inhibitor in mice. I. Increase during infection, *Immunology*, 56, 113, 1985.

71. **Playfair, J. H. L., DeSouza, J. B., and Cottrell, B. J.**, Reactivity and cross-reactivity of mouse helper T cells to malaria parasites, *Immunology*, 32, 681, 1977.

72. **Kabilan, L., Troye-Blomberg, M., Patarroyo, M. E., Bjorkman, A., and Perlmann, P.**, Regulation of the immune response in *P. falciparum* malaria. IV. T cell dependent production of immunoglobulin and anti-*P. falciparum* antibodies *in vitro*, *Clin. Exp. Immunol.*, 68, 288, 1987.

73. **Nardin, E. H., Barr, P. J., Gibson, H. L., Collins, W. E., Nussenzweig, R. S., and Nussenweig, V.**, Induction of sporozoite-specific memory cells in mice immunized with a recombinant *Plasmodium vivax* circumsporozoite protein, *Eur. J. Immunol.*, 17, 1763, 1987.

74. **McDonald, V. and Phillips, R. S.**, *Plasmodium chabaudi:* adoptive transfer of immunity with different spleen cell populations and development of protective activity in the serum of lethally irradiated recipient mice, *Exp. Parasitol.*, 49, 26, 1980.

75. **Strickland, G. T., Ahmed, A., Sayles, P. C., and Hunter, K. W.**, Murine malaria: cellular interactions in the immune response, *Am. J. Trop. Med. Hyg.*, 32, 1229, 1983.

76. **McDonald, V. and Phillips, R. S.**, *Plasmodium chabaudi* in mice. Adoptive transfer of immunity with enriched populations of spleen T and B lymphocytes, *Immunology*, 34, 821, 1978.

77. **Gravely, S. M. and Kreier, J. P.**, Adoptive transfer of immunity to *Plasmodium berghei* with immune T and B lymphocytes, *Infect. Immun.*, 14, 184, 1976.

78. **Brown, K. N., Jarra, W., and Hills, L. A.**, T cells and protective immunity to *Plasmodium berghei* with immune T and B lymphocytes, *Infect. Immun.*, 14, 858.

79. **Brown, K. N., Hills, L. A., and Jarra, W.**, Preliminary studies of artificial immunization of rats against *Plasmodium berghei* and adoptive transfer of this immunity by splenic T and T + B cells, *Bull. W.H.O.*, 54, 149, 1976.

80. **Cavacini, L. A., Long, C. A., and Weidanz, W. P.**, T-cell immunity in murine malaria: adoptive transfer of resistance to *Plasmodium chabaudi adami* in nude mice with splenic T cells, *Infect. Immun.*, 52, 637, 1986.

81. **Fahey, J. R. and Spitalny, G. L.**, Immunity to *Plasmodium yoelii*: kinetics of the generation of T and B lymphocytes that passively transfer protective immunity against virulent challenge, *Cell. Immunol.*, 98, 486, 1986.

82. **Zuckerman, A. and Jacobson, R. L.**, Adoptive transfer of immunity to *Plasmodium berghei* by a population of immune rat spleen cells resistant to cyclophosphamide, *Int. J. Parasitol.*, 6, 103, 1976.

83. **Phillips, R. S.**, *Plasmodium berghei:* passive transfer of immunity by antisera and cells, *Exp. Parasitol.*, 27, 479, 1970.

84. **Mogil, R. J., Patton, C. L., and Green, D. R.**, Cellular subsets involved in cell-mediated immunity to murine *Plasmodium yoelii* 17X malaria, *J. Immunol.*, 138, 1933, 1987.

85. **Brinkmann, V., Kaufmann, S. H. E., and Simon, M. M.**, T-cell-mediated immune response in murine malaria: differential effects of antigen-specific Ly T-cell subsets in recovery from *Plasmodium yoelii* infection in normal and T-cell-deficient mice, *Infect. Immun.*, 47, 737, 1985.

86. **Habu, S., Yamauchi, K., Gershon, R. K., and Murphy, D. B.**, A non-T:non-B cell bears I-A, I-E, I-J and T1A (Qa-1?) determinants, *Immunogenetics*, 13, 215, 1981.

87. **Murphy, D. B., Yamauchi, K., Habu, S., Eardley, D. D., and Gershon, R. K.**, T cells in a suppressor circuit and non-T:non-B cells bear different I-J determinants, *Immunogenetics*, 13, 205, 1981.

88. **Verhave, J. P., Strickland, G. T., Jaffe, H. A., and Ahmed, A.**, Studies on the transfer of protective immunity with lymphoid cells from mice immune to malaria sporozoites, *J. Immunol.*, 121, 1031, 1978.

89. **Egan, J. E., Weber, J. L., Ballou, W. R., Hollingdale, M. R., Majarian, W. R., Gordon, D. M., Maloy, W. L., Hoffman, S. L., Wirtz, R. A., Schneider, I., Woollett, G. R., Young, J. F., and Hockmeyer, W. T.**, Efficacy of murine malaria sporozoite vaccines: implications for human vaccine development, *Science*, 236, 453, 1987.

90. **Harte, P. G., Rogers, N. C., and Targett, G. A. T.**, Role of T cells in preventing transmission of rodent malaria, *Immunology*, 56, 1, 1985.

91. **Gross, A., Frankenburg, S., and Londner, M. V.**, Cell-mediated immunity in rats injected with an antimalaria T-cell line, *Cell. Immunol.*, 84, 14, 1984.

92. **Chemtai, A. K., Vaeck, M., Hamers-Casterman, C., Hamers, R., and De Baetselier, P.**, T-cell mediated immunity in murine malaria. I. Induction of T-cell dependent proliferative responses to *Plasmodium chabaudi*, *Parasite Immunol.*, 6, 51, 1984.

93. **Chemtai, A. K., Hamers-Casterman, C., Hamers, R., and De Baestselier, P.**, T cell-mediated immunity in murine malaria. II. Induction of protective immunity to *P. chabaudi* by antigen fed macrophages and antigen educated lymphocytes, *Parasite Immunol.*, 6, 469, 1984.

94. **Sinigaglia, F., Matile, H., and Pink, J. R. L.**, *Plasmodium falciparum* -specific human T cell clones: evidence for helper and cytotoxic activities, *Eur. J. Immunol.*, 17, 187, 1987.

95. **Pink, J. R. L., Rijnbeek, A.-M., Reber-Liske, R., and Sinigaglia, F.**, *Plasmodium falciparum* -specific human T cell clones: recognition of different parasite antigens, *Eur. J. Immunol.*, 17, 193, 1987.

96. **Brake, D. A., Long, C. A., and Weidanz, W. P.**, Adoptive protection against *Plasmodium chabaudi adami* malaria in athymic nude mice by a cloned T cell line, *J. Immunol.*, 140, 1989, 1988.

97. **Good, M. F., Quakyi, I. A., Saul, A., Berzofsky, J. A., Carter, R., and Miller, L. H.**, Human T clones reactive to the sexual stages of *Plasmodium falciparum* malaria. High frequency of gamete-reactive T cells in peripheral blood from nonexposed donors, *J. Immunol.*, 138, 306, 1987.

98. **Chizzolini, C. and Perrin, L.**, Antigen-specific and MHC-restricted *Plasmodium falciparum*-induced human T lymphocytes clones, *J. Immunol.*, 137, 1022, 1986.

99. **Simitsek, P., Chizzolini, C., and Perrin, L.**, Malaria specific human T cell clones: crossreactivity with various plasmodia species, *Clin. Exp. Immunol.*, 69, 271, 1987.

100. **Carlin, J. M., Jensen, J. B., and Geary, T. G.**, Comparison of inducers of crisis forms in *Plasmodium falciparum in vitro*, *Am. J. Trop. Med. Hyg.*, 34, 668, 1985.

101. **Clark, I. A., Hunt, N. H., Butcher, G. A., and Cowden, W. B.**, Inhibition of murine malaria (*Plasmodium chabaudi*) *in vivo* by recombinant interferon-γ or tumor necrosis factor, and its enhancement by butylated hydroxyanisole, *J. Immunol.*, 139, 3493, 1987.

102. **Playfair, J. H. L. and DeSouza, J. B.**, Recombinant gamma interferon is a potent adjuvant for a malaria vaccine in mice, *Clin. Exp. Immunol.*, 67, 5, 1987.

103. **Jayawardena, A. N., Mogil, R., Murphy, D. B., Burger, D., and Gershon, R. K.**, Enhanced expression of H2-K and H2-D antigens on reticulocytes infected with *Plasmodium yoelii*, *Nature*, 302, 623, 1983.

104. **Coleman, R. M., Prencricca, N. H., Stout, J. P., Brisette, W. H., and Smith, D. M.**, Splenic mediated erythrocyte cytotoxicity in malaria, *Immunology*, 29, 49, 1975.

105. **Orago, A. S. S. and Solomon, J. B.**, Antibody-dependent and -independent cytotoxic activity of spleen cells for *Plasmodium berghei* from susceptible and resistant rats, *Immunology*, 59, 283, 1986.

106. **Schofield, L., Villaquiran, J., Ferreira, A., Schellekens, H., Nussenzweig, R. S., and Nussenzweig, V.**, Interferon, CD8+ T cells and antibodies required for immunity to malaria sporozoites, *Nature*, 330, 664, 1987.

107. **Weiss, W. R., Sedegah, M., Beaudoin, R. L., Miller, L. H., and Good, M. F.**, CD8+ T cells (cytotoxic/suppressors) are required for protection in mice immunized with malaria sporozoites, *Proc. Natl. Acad. Sci. U.S.A.*, 85, 573, 1988..

108. **Ferreira, A., Schofield, L., Enea, V., Schellekens, H., Van Der Meide, P., Collins, W. E., Nussenzweig, R. S., and Nussenzweig, V.**, Inhibition of development of exoerythrocytic forms of malaria parasites by γ-interferon, *Science*, 232, 881, 1986.

109. **Maheshwari, R. K., Czarniecki, C. W., Dutta, G. P., Puri, S. K., Dhawan, B. N., and Friedman, R. M.**, Recombinant human gamma interferon inhibits simian malaria, *Infect. Immun.*, 53, 628, 1986.

110. **Mellouk, S., Maheshwari, R. K., Rhodes-Feuillette, A., Beaudoin, R. L., Berbiguier, N., Matile, H., Miltgen, F., Landau, I., Pied, S., Chigot, J. P., Friedman, R. M., and Mazier, D.**, Inhibitory activity of interferons and interleukin 1 on the development of *Plasmodium falciparum* in human hepatocyte cultures, *J. Immunol.*, 139, 4192, 1987.

111. **Schofield, L., Ferreira, A., Altzuler, R., Nussenzweig, V., and Nussenzweig, R. S.**, Interferon-γ inhibits the intrahepatocytic development of malaria parasites *in vitro*, *J. Immunol.*, 139, 2020, 1987.

112. **Ballou, W. R., Rothbard, J., Wirtz, R. A., Gordon, D. M., Williams, J. S., Gore, R. W., Schneider, I., Hollingdale, M. R., Beaudoin, R. L., Maloy, W. L., Miller, L. H., and Hockmeyer, W. T.**, Immunogenicity of synthetic peptides from circumsporozoite protein of *Plasmodium falciparum*, *Science*, 228, 996, 1985.

113. **Mellouk, S., Mazier, D., Druilhe, P., Berbiguier, N., and Danis, M.**, *In vitro* and *in vivo* results suggest that anti sporozoite antibodies do not totally block *Plasmodium falciparum* sporozoite infectivity, *N. Engl. J. Med.*, 315, 648, 1986.

114. **Ruebush, M. J. and Burgess, D. E.**, Induction of natural killer cells and interferon production during infection of mice with *Babesia microti* of human origin, in *NK Cells and Other Natural Effector Cells*, Herberman, R. B., Ed., Academic Press, New York, 1982, 1483.

115. **Eugui, E. M. and Allison, A. C.**, Differences in susceptibility of various mouse strains to haemoprotozoa infections: possible correlation with natural killer activity, *Parasite Immunol.*, 2, 277, 1980.

116. **Troye-Blomberg, M., Andersson, G., Stoczkowska, M., Shabo, R., Romero, P., Patarroyo, E., Wigzell, H., and Perlmann, P.**, Production of IL 2 and IFN-γ by T cells from malaria patients in response to *Plasmodium falciparum* or erythrocyte antigens *in vitro*, *J. Immunol.*, 1356, 3498, 1985.

117. **Brunda, M. J. and Davatelis, V.**, Augmentation of natural killer cell activity by recombinant interleukin-2 and recombinant interferons, in *Mechanisms of Cytotoxicity by NK Cells*, Herberman, R. B. and Callewaert, D. M., Eds., Academic Press, New York, 1985, 397.

118. **Braakman, E., Van Tunen, A., Meager, A., and Lucas, C. J.**, IL-2 and IFN-γ enhanced natural cytotoxicity: analysis of the role of different lymphoid subsets and implications for activation routes, *Cell. Immunol.*, 99, 476, 1986.

119. **Stach, J. l., Dufrenoy, E., Roffi, J., and Bach, M. A.**, T-cell subsets and natural killer activity in *Plasmodium falciparum*-infected children, *Clin. Immunol. Immunopathol.*, 38, 129, 1986.

120. **Ojo-Amaize, E. A., Salimonu, L. S., Williams, A. I. O., Akinwolere, O. A. O., Shabo, R., Alm, G. V., and Wigzell, H.,** Positive correlation between degree of parasitemia, interferon titers, and natural killer cell activity in *Plasmodium falciparum*-infected children, *J. Immunol.,* 127, 2296, 1981.

121. **Ojo-Amaize, E. A., Vilcek, J., Cochrane, A. H., and Nussenzweig, R. S.,** *Plasmodiun berghei* sporozoites are mitogenic for murine T cells, induce interferon, and activate natural killer cells, *J. Immunol.,* 133, 1005, 1984.

122. **Hunter, K. W., Jr., Folks, T. M., Sayles, P. C., and Strickland, G. T.,** Early enhancement followed by suppression of natural killer cell activity during murine malarial infections, *Immunol. Lett.,* 2, 209, 1981.

123. **Skamene, E., Stevenson, M. M., and Lemieux, S.,** Murine malaria: dissociation of natural killer (NK) cell activity and resistance to *Plasmodium chabaudi, Parasite Immunol.,* 5, 557, 1983.

124. **Wood, P. R. and Clark, I. A.,** Apparent irrelevance of NK cells to resolution of infections with *Babesia microti* and *Plasmodium vinckei petteri* in mice, *Parasite Immunol.,* 4, 319, 1982.

125. **Kindred, B.,** Nude mice in immunology, *Prog. Allergy,* 26, 137, 1979.

126. **Kharazmi, A. and Jepsen, S.,** Enhanced inhibition of *in vitro* multiplication of *Plasmodium falciparum* by stimulated human polymorphonuclear leucocytes, *Clin. Exp. Immunol.,* 57, 287, 1984.

127. **Trubowitz, S. and Masek, B.,** *Plasmodium falciparum:* phagocytosis by polymorphonuclear leucocytes, *Science,* 162, 273, 1968.

128. **Tosta, C. F. and Wedderburn, N.,** Immune phagocytosis of *Plasmodium yoelii*-infected erythrocytes by macrophages and eosinophils, *Clin. Exp. Immunol.,* 42, 114, 1980.

129. **Celada, A., Cruchaud, A., and Perrin, L. H.,** Phagocytosis of *Plasmodium falciparum*-parasitized erythrocytes by human polymorphonuclear leukocytes, *J. Parasitol.,* 69, 59, 1983.

130. **Brown, J. and Smalley, M. E.,** Inhibition of the *in vitro* growth of *Plasmodium falciparum* by human polymorphonuclear neutrophil leucocytes, *Clin. Exp. Immunol.,* 46, 106, 1981.

131. **Sinden, R. E. and Smalley, M. E.,** Gametocytes of *Plasmodium falciparum:* phagocytosis by leucocytes *in vivo* and *in vitro, Trans. R. Soc. Trop. Med. Hyg.,* 10, 344, 1976.

132. **Nnalue, N. A. and Friedman, M. J.,** Evidence for a neutrophil-mediated protective response in malaria, *Parasite Immunol.,* 10, 47, 1988.

133. **Waters, L. S., Taverne, J., Tai, P.-C., Spry, C. J. F., Targett, G. A. T., and Playfair, J. H. L.,** Killing of *Plasmodium falciparum* by eosinophil secretory products, *Infect. Immun.,* 55, 877, 1987.

134. **Maury, C. P. J.,** Tumour necrosis factor-An overview, *Acta Med. Scand.,* 220, 387, 1986.

135. **Taliaferro, W. H. and Mulligan, H. W.,** The histopathology of malaria with special reference to the function and origin of macrophages in defence, *Indian Med. Res. Mem.,* 29, 1, 1937.

136. **Lee, S.-H., Crocker, P., and Gordon, S.,** Macrophage plasma membrane and secretory properties in murine malaria. Effects of *Plasmodium yoelii* blood-stage infection on macrophages in liver, spleen, and blood, *J. Exp. Med.,* 163, 54, 1986.

137. **Shear, H. L., Nussenzweig, R. S., and Bianco, C.,** Immune phagocytosis in murine malaria, *J. Exp. Med.,* 149, 1288, 1979.

138. **Brown, J., Greenwood, B. M., and Terry, R. J.,** Cellular mechanisms involved in recovery from acute malaria in Gambian children, *Parasite Immunol.,* 8, 551, 1986.

139. **Cantrell, W., Elko, E. E., and Hopff, B. M.,** *Plasmodium berghei:* phagocytic hyperactivity of infected rats, *Exp. Parasitol.,* 28, 291, 1970.

140. **Lucia, H. L. and Nussenzweig, R. S.,** *Plasmodium chabaudi* and *Plasmodium vinckeii:* phagocytic activity of mouse reticuloendothelial system, *Exp. Parasitol.,* 25, 319, 1969.

141. **Playfair, J. H. L.,** Lethal *Plasmodium yoelii* malaria: the role of macrophages in normal and immunized mice, *Bull. W.H.O.,* 57, (Suppl. 1), 245, 1979.

142. **Allison, A. C. and Eugui, E. M.,** A radical interpretation of immunity to malaria parasites, *Lancet,* ii, 431, 1982.

143. **Clark, I. A. and Hunt, N. H.,** Evidence for reactive oxygen intermediates causing hemolysis and parasite death in malaria, *Infect. Immun.,* 39, 1, 1983.

144. **Ockenhouse, C. F. and Shear, H. L.,** Malaria-induced lymphokines: stimulation of macrophages for enhanced phagocytosis, *Infect. Immun.,* 42, 733, 1983.

145. **Ockenhouse, C. F. and Shear, H. L.,** Oxidative killing of the intraerythrocytic malaria parasite *Plasmodium yoelii* by activated macrophages, *J. Immunol.,* 132, 424, 1984.

146. **Ockenhouse, C. F., Schulman, S., and Shear, H. L.,** Induction of crisis forms in the human malaria parasite *Plasmodium falciparum* by γ-interferon-activated, monocyte-derived macrophages, *J. Immunol.,* 133, 1601, 1984.

147. **Wyler, D. J., Herrod, H. G., and Weinbaum, F. I.,** Response of sensitized and unsensitized human lymphocyte subpopulations to *Plasmodium falciparum* antigens, *Infect. Immun.,* 24, 106, 1979.

148. **Khusmith, S., Druilhe, P., and Gentilini, M.,** Enhanced *Plasmodium falciparum* merozoite phagocytosis by monocytes from immune individuals, *Infect. Immun.,* 35, 874, 1982.

149. **Chow, J. W. and Kreier, J. F,** *Plasmodium berghei:* adherence and phagocytosis by rat macrophages *in vitro, Exp. Parasitol.,* 31, 13, 1972.

150. **Shear, H. L.**, Murine malaria: immune complexes inhibit Fc receptor-mediated phagocytosis, *Infect. Immun.*, 44, 130, 1984.

151. **Haidaris, C. G., Haynes, D., Meltzer, M. S., and Allison, A. C.**, Serum containing tumor necrosis factor is cytotoxic for the human malaria parasite *Plasmodium falciparum, Infect. Immun.*, 42, 385, 1983.

152. **Taliaferro, W. H. and Taliaferro, L. G.**, The effect of immunity on the asexual reproduction of *Plasmodium brasilianeum, J. Infect. Dis.*, 75, 1, 1944.

153. **Clark, I. A., Richmond, J. E., Willis, E. J., and Allison, A. C.**, Intra-erythrocytic death of the parasite in mice recovering from infection with *Babesia microti, Parasitology*, 75, 189, 1977.

154. **Friedman, M. J.**, Oxidant damage mediates variant red cell resistance to malaria, *Nature*, 280, 245, 1979.

155. **Eaton, J. W., Eckman, J. R., Berger, E., and Jacob, H. S.**, Suppression of malaria infection by oxidant-sensitive host erythrocytes, *Nature*, 264, 758, 1976.

156. **Eckman, J. R. and Eaton, J. W.**, Dependence of plasmodial glutathione metabolism on the host cell, *Nature*, 378, 754. 1979.

157. **Wozencraft, A. O., Croft, S. L., and Sayers, G.**, Oxygen radical release by adherent cell populations during the initial stages of a lethal rodent malarial infection, *Immunology*, 56, 523, 1985.

158. **Dockrell, H. M., Alavi, A., and Playfair, J. H. L.**, Changes in oxidative burst capacity during murine malaria and the effect of vaccination, *Clin. Exp. Immunol.*, 66, 37, 1986.

159. **Brinkmann, V., Kaufmann, S. H. E., Simon, M. M., and Fischer, H.**, Role of macrophages in malaria; O_2^- metabolite production and phagocytosis by splenic macrophages during lethal *Plasmodium berghei* and self-limiting *Plasmodium yoelii* infection in mice, *Infect. Immun.*, 44, 743, 1984.

160. **Descamps-Latscha, B., Lunel-Fabiani, F., Kara-Binis, A., and Druilhe, P.**, Generation of reactive oxygen species in whole blood from patients with acute falciparum malaria, *Parasite Immunol.*, 9, 275, 1987.

161. **Wozencraft, A. O., Dockrell, H. M., Taverne, J., Targett, G. A. T., and Playfair, J. H. L.**, Killing of human malaria parasites by macrophages secretory products, *Infect. Immun.*, 43, 664, 1984.

162. **Dockrell, H. M. and Playfair, J. H. L.**, Killing of *Plasmodium yoelii* by enzyme-induced products of the oxidative burst, *Infect. Immun.*, 43, 451, 1984.

163. **Dockrell, H. M. and Playfair, J. H. L.**, Killing of blood-stage murine malaria parasites by hydrogen peroxide, *Infect. Immun.*, 39, 456, 1983.

164. **Rigdon, R. H., Micks, D. W., and Breslin, D.**, Effect of phenylhydrazine hydrochloride on *Plasmodium knowlesi* infection in the monkey, *Am. J. Hyg.*, 52, 308, 1950.

165. **Clark, I. A., Cowden, W. B., and Butcher, G. A.**, Free oxygen radical generators as antimalarial drugs, *Lancet*, i, 234, 1983.

166. **Clark, I. A., Hunt, N. H., Cowden, W. B., Maxwell, L. E., and Mackie, E. J.**, Radical-mediated damage to parasites and erythrocytes in *Plasmodium vinckei* infected mice after injection of t-butyl hydroperoxide, *Clin. Exp. Immunol.*, 56, 524, 1984.

167. **Clark, I. A., Cowden, W. B., Hunt, N. H., Maxwell, L. E., and Mackie, E. J.**, Activity of divicine in *Plasmodium vinckei*-infected mice has implications for treatment of favism and epidemiology of G-6-PD deficiency, *Br. J. Haematol.*, 57, 479, 1984.

168. **Gutteridge, J. M. C., Richmond, R., and Halliwell, B.**, Inhibition of the iron-catalyzed formation of hydroxyl radicals from superoxide and of lipid peroxidation by desferrioxamine, *Biochem. J.*, 184, 469, 1979.

169. **Stocker, R., Hunt, N. H., Buffinton, G. D., Weidemann, M. J., Lewis-Hughes, P. H., and Clark, I. A.**, Oxidative stress and protective mechanisms in erythrocytes in relation to *Plasmodium vinckei* load, *Proc. Natl. Acad. Sci. U.S.A.* 82, 548, 1985.

170. **Lynch, R. E. and Fridovich, I.**, Permeation of the erythrocyte stroma by superoxide radical, *J. Biol. Chem.*, 253, 4697, 1978.

171. **Jain, S. K., Mohandas, N., Clark, M. R., and Shohet, S. B.**, The effect of malonyldialdehyde, a product of lipid peroxidation, on the deformability, dehydration and ^{51}Cr-survival of erythrocytes, *Br. J. Haematol.*, 53, 247, 1983.

172. **Clark, I. A., Butcher, G. A., Buffinton, G. D., Hunt, N. H., and Cowden, W. B.**, Toxicity of certain products of lipid peroxidation to the human malaria parasite *Plasmodium falciparum, Biochem. Pharmacol.*, 36, 543, 1987.

173. **Buffinton, G. D., Hunt, N. H., Cowden, W. B., and Clark, I. A.**, Detection of short-chain carbonyl products of lipid peroxidation from malaria-parasite (*Plasmodium vinckei*)-infected red blood cells exposed to oxidative stress, *Biochem. J.*, 249, 63, 1988.

174. **Buffinton, G. D., Hunt, N. H., Cowden, W. B., and Clark, I. A.**, Cell damage and disease, in *Free Radicals*, Rice-Evans, C., Ed., Richelieu Press, London, 1986, 201,

175. **Kharazmi, A., Jepsen, S., and Valerius, N. H.**, Polymorphonuclear leucocytes defective in oxidative metabolism inhibit *in vitro* growth of *Plasmodium falciparum*. Evidence against an oxygen-dependent mechanism, *Scand. J. Immunol.*, 20, 93, 1984.

176. **Mannel, D. N., Moore, R. N., and Mergenhagen, S. E.**, Macrophages as a source of tumoricidal activity (tumor-necrotizing factor), *Infect. Immun.*, 30, 523, 1980.

177. **Taverne, J., Dockrell, H. M., and Playfair, J. H. L.**, Endotoxin-induced serum factor kills malarial parasites *in vitro*, *Infect. Immun.*, 33, 83, 1981.

178. **Taverne, J., Depledge, P., and Playfair, J. H. L.**, Differential sensitivity *in vivo* of lethal and nonlethal malarial parasites to endotoxin-induced serum factor, *Infect. Immun.*, 37, 927, 1982.

179. **Taverne, J., Matthews, N., Depledge, P., and Playfair, J. H. L.**, Malarial parasites and tumour cells are killed by the same component of tumour necrosis serum, *Clin. Exp. Immunol.*, 57, 293, 1984.

180. **Clark, I. A.**, Does endotoxin cause both the disease and parasite death in acute malaria and babesiosis?, *Lancet*, ii, 75, 1978.

181. **Wood, P. R. and Clark, I. A.**, Macrophages from *Babesia* and malaria infected mice are primed for monokine release, *Parasite Immunol.*, 6, 309, 1984.

182. **Clark, I. A., Virelizier, J.-L., Carswell, E. A., and Wood, P.**, Possible importance of macrophage-derived mediators in acute malaria, *Infect. Immun.*, 32, 1058, 1981.

183. **Rzepczyk, C. M. and Clark, I. A.**, Demonstration of a lipopolysaccharide-induced cytostatic effect on malarial parasites, *Infect. Immun.*, 33, 343, 1981.

184. **Rzepczyk, C. M. and Clark, I. A.**, Failure of bacterial lipopolysaccharide to elicit a cytostatic effect on *Plasmodium vinckei petteri* in C3H/HeJ mice, *Infect. Immun.*, 35, 58, 1982.

185. **Tubbs, H.**, Endotoxin in human and murine malaria, *Trans. R. Soc. Trop. Med. Hyg.*, 74, 121, 1980.

186. **Scuderi, P., Lam, K. S., Ryan, K. J., Petersen, E., Sterling, K. E., Finley, P. R., Ray, C. G., Slymen, D. J., and Salmon, S. E.**, Raised levels of tumour necrosis factor in parasitic infections, *Lancet*, ii, 1364, 1986.

187. **Nedwin, G. E., Svedersky, L. P., Bringman, T. S., Palladino, M. A., Jr., and Goeddel, D. V.**, Effect of interleukin 2, interferon-γ and mitogens on the production of tumour necrosis factors α and β, *J. Immunol.*, 135, 2492, 1985.

188. **Jensen, J. B., Vande Waa, J. A., and Karadsheh, A. J.**, Tumor necrosis factor does not induce *Plasmodium falciparum* crisis forms, *Infect. Immun.*, 55, 1722, 1987.

189. **Cavacini, L. A. and Weidanz, W. P.**, unpublished observations, 1988.

190. **Taverne, J., Tavernier, J., Fiers, W., and Playfair, J. H. L.**, Recombinant tumour necrosis factor inhibits malaria parasites *in vivo* but not *in vitro*, *Clin. Exp. Immunol.*, 67, 1, 1987.

191. **Silberstein, D. and David, J. R.**, Tumor necrosis factor enhances eosinophil toxicity to *Schistosoma mansoni* larvae, *Proc. Natl. Acad. Sci., U.S.A.*, 83, 1055, 1986.

192. **Thorne, K. J. I., Richardson, B. A., Taverne, J., Williamson, D. J., Vadas, M. A., and Butterworth, A. E.**, A comparison of eosinophil-activating factor (EAF) with other monokines and lymphokines, *Eur. J. Immunol.*, 16, 1143, 1986.

193. **Shalaby, M. R., Aggarwal, B. B., Rinderknecht, E., Svedersky, L. P., Finkle, B. S., and Palladino, M. A., Jr.**, Activation of human polymorphonuclear neutrophil functions by interferon-γ and tumor necrosis factors, *J. Immunol.*, 135, 2069, 1985.

194. **Tappel, A. L.**, Lipid peroxidation and fluorescent molecular damage to membranes, in *Pathobiology of Cell Membranes*, Vol. 1, Trump, B. F. and Arstila, F., Eds., Academic Press, New York, 1975, 311.

195. **Egan, J. E., Haynes, J. D., Brown, N. D., and Eisemann, C. S.**, Polyamine oxidase in human retroplacental serum inhibits the growth of *Plasmodium falciparum*, *Am. J. Trop. Med. Hyg.*, 35, 890, 1986.

196. **Jensen, J. B., Boland, M. T., Allan, J. S., Carlin, J. M., Vande Waa, J. A., Divo, A. A., and Akood, M. A. S.**, Association between human serum-induced crisis forms in cultured *Plasmodium falciparum* and clinical immunity to malaria in Sudan, *Infect. Immun.*, 41, 1302, 1983.

197. **Jensen, J. B., Boland, M. T., and Akood, M. A. S.**, Induction of crisis forms in cultured *Plasmodium falciparum* with human serum from Sudan, *Science*, 216, 1230, 1982.

198. **Wyler, D. J., Oster, C. N., and Quinn, T. C.**, The role of the spleen in malaria infections, in *The Role of the Spleen in the Immunology of Parasitic Diseases*, Tropical Diseases Research Series, No. 1, Torrigiani, G., Ed., Scwabe & Co. AG, Basel, 1979, 1983.

199. **Weiss, L.**, The spleen, in *Histology, Cell and Tissue Biology*, 5th ed., Weiss, L., Ed., Elsevier, New York, 1983, 544.

200. **Roberts, D. W. and Weidanz, W. P.**, Splenomegaly, enhanced phagocytosis, and anemia are thymus dependent responses to malaria, *Infect. Immun.*, 20, 728, 1978.

201. **Playfair, J. H. L. and DeSouza, J. B.**, Lymphocyte traffic and lymphocyte destruction in murine malaria, *Immunology*, 46, 125, 1982.

202. **Kumararathe, D. S., Phillips, R. S., Sinclair, D., Parrott, M. V. D., and Forrester, J. B.**, Lymphocyte migration in murine malaria during the primary patent parasitemia of *Plasmodium chabaudi* infections, *Clin. Exp. Immunol.*, 68, 65, 1987.

203. **Taverne, J., Rahman, D., Dockrell, H. M., Alavi, A., Leveton, C., and Playfair, J. H. L.**, Activation of liver macrophages in murine malaria is enhanced by vaccination, *Clin. Exp. Immunol.*, 70, 508, 1987.

204. **Grun, J. L., Long, C. A., and Weidanz, W. P.**, Effects of splenectomy on antibody-independent immunity to *Plasmodium chabaudi adami* malaria, *Infect. Immun.*, 48, 853, 1985.

205. **Moran, C. J., De Rivera, V. S., and Turk, J. L.**, The immunological significance of histological changes in the spleen and liver in mouse malaria, *Clin. Exp. Immunol.*, 42, 412, 1973.

206. **Silverman, P. H., Schooley, J. C., and Mahlmann, L. J.,** Murine malaria decreases hemopoietic stem cells, *Blood,* 69, 408, 1987.

207. **Wyler, D. J., Quinn, T. C., and Chen, L. T.,** Relationship of alterations in splenic clearance function and microcirculation to host defense in acute rodent malaria, *J. Clin. Invest.,* 67, 1400, 1981.

208. **Weiss, L., Geduldig, U., and Weidanz, W. P.,** Mechanisms of splenic control of murine malaria: reticular cell activation and the development of a blood-spleen barrier, *Am. J. Anat.,* 176, 251, 1986.

209. **Good, M. F., Berzofsky, J. A., Maloy, W. L., Hayashi, Y., Fujii, N., Hockmeyer, W. T., and Miller, L. H.,** Genetic control of the immune response in mice to a *Plasmodium falciparum* sporozoite vaccine. Widespread nonresponsiveness to single malaria T epitope in highly repetitive vaccine, *J. Exp. Med.,* 164, 655, 1986.

210. **Del Giudice, G., Cooper, J. A., Merino, J., Verdini, A. S., Pessi, A., Togna, A. R., Engers, H. D., Corradin, G., and Lambert, P.-H.,** The antibody response in mice to carrier-free synthetic polymers of *Plasmodium falciparum* circumsporozoite repetitive epitope is I-A^b-restricted: possible implications for malaria vaccines, *J. Immunol.,* 137, 2952, 1986.

211. **Togna, A. R., Del Giudice, G., Verdini, A. S., Bonelli, F., Pessi, A., Engers, H. D., and Corradin, G.,** Synthetic *Plasmodium falciparum* circumsporozoite peptides elicit heterogeneous L3T4$^+$ T cell proliferative responses in H-2^b mice, *J. Immunol.,* 137, 2956, 1986.

212. **Good, M. F., Pombo, D., Quakyi, I. A., Riley, E. M., Houghten, R. A., Menon, A., Alling, D. W., Berzofsky, J. A., and Miller, L. H.,** Human T-cell recognition of the circumsporozoite protein of *Plasmodium falciparum*: immunodominant T-cell domains map to the polymorphic regions of the molecule, *Proc. Natl. Acad. Sci., U.S.A.,* 85, 1199, 1988.

213. **Nussenzweig, R. S. and Nussenzweig, V.,** Development of sporozoite vaccines, *Trans. R. Soc. Lond. B,* 307, 117, 1984.

214. **Playfair, J. H. L., DeSouza, J. B., Freeman, R. R., and Holder, A. A.,** Vaccination with a purified blood-stage malaria in mice: correlation of protection with T cell mediated immunity, *Clin. Exp. Immunol.,* 62, 19, 1985.

215. **Good, M. F., Maloy, W. L., Lunde, M. N., Margalit, H., Cornette, J. L., Smith, G. L., Moss, B., Miller, L. H., and Berzofsky, J. A.,** Construction of synthetic immunogen: use of new T-helper epitope on malaria circumsporozoite protein, *Science,* 235, 1059, 1987.

Chapter 3

CELL-MEDIATED IMMUNITY AND ITS ROLE IN PROTECTION

John H. L. Playfair, K. Rebecca Jones, and Janice Taverne

TABLE OF CONTENTS

I. INTRODUCTION

This chapter is principally concerned with cell-mediated immunity (CMI) in the sense of nonantibody effector mechanisms and their modification by T cells and T cell products. Our interest in this topic began some 12 years ago when we found that, in a murine blood-stage malaria model, protection by vaccination correlated better with various cell-mediated changes than with antibody levels.[1] We subsequently showed that parasites could be killed in several ways by myeloid cells and their products, some of which could be enhanced by vaccination; this work is reviewed in Section IV.

However, the idea of nonantibody-dependent immunity in malaria goes back at least to the 1940s when Taliaferro noted that during the "crisis" phase of blood-stage malaria degenerating parasites could be seen within red cells, and concluded that they were being killed, or at least damaged, before the infected red cells were removed by phagocytosis and not after, as had been previously thought.[2] Interest then swung towards antibody, which was shown in the classic experiment of Cohen et al.[3] to reduce parasitemia upon transfer to children infected with *Plasmodium falciparum*. The development of monoclonal antibodies gave further impetus to this line of research because of their great power in selecting the antigens which antibodies recognize.

At about the same time, Clark and colleagues[4,5] revived interest in the ideas of Taliaferro by demonstrating that various nonspecific stimuli, such as BCG and *Corynebacterium parvum* (now *Propionobacterium acnes*) could induce both immunity and crisis forms in mice infected with *Plasmodium* or *Babesiae*. Further support for effective nonantibody immunity came from the work of Grun and Weidanz[6,7] with B cell deficient mice, which develop normal immunity to *P. chabaudi adami* and, if drug cured of the primary infection, to *P. yoelii*. This immunity involved the spleen and required T cells, and would clearly fall into the category of cell-mediated, though it is not yet clear what the actual effector mechanism is. Meanwhile, a number of cell products, notably of the macrophage, were found to be cytotoxic to blood-stage parasites (see Section II). A recent interesting development has been the discovery that the liver stage too is susceptible to nonantibody cytotoxicity (see Section III).

In this review we will consider first the molecules, other than antibody, which have been shown to be actually or potentially cytotoxic to malaria parasites, then the role of various cytokines which, while not directly cytotoxic, enhance or regulate cytotoxic mechanisms, and finally the evidence that T cells are involved in such enhancement and that further enhancement may be also induced by vaccines.

II. PARASITE KILLING MECHANISMS

A. OXIDATIVE
1. Susceptibility of Parasitized Red Cells

Blood stage malaria parasites are known to render erythrocytes more susceptible to oxidant stress,[8] perhaps because parasite-derived reactive oxygen intermediates (ROI) are generated within the red cell which cause damage by peroxidation of membrane lipids.[9] The parasite has been shown to modulate many of the mechanisms that protect the cell against oxygen stress, such as glutathione peroxidase, superoxide dismutase, catalase, and hemo-globin,[10,11] and this may explain why cultivation of *P. falciparum in vitro* is only possible in conditions of reduced oxygen tension.[12] A parasite within a red cell exposed to oxidant stress may itself be damaged as a result of glutathione oxidation and lipid peroxidation of the red cell membrane, and in extreme cases a net efflux of potassium ions may cause the red cell to lyse.[13,14]

Damage to the red cell membrane, in addition to any direct effects on the parasite within, may cause changes in membrane architecture and hydrophilicity which prevent invasion of parasites and which might be responsible for the well-known removal of both uninfected and infected red cells from the circulation.

2. Hydrogen Peroxide

Evidence from several groups, including our own, suggests that hydrogen peroxide is not only toxic to *P. falciparum*,[15] *P. yoelii*, and *P. berghei*[16] *in vitro* when freshly generated in cell-free enzyme systems, but also when released extracellularly from effector cells.[17] Antibody-independent killing of malaria parasites by macrophages is at least partially reversible by catalase, which detoxifies hydrogen peroxide.[15,18]

3. Hydroxyl Radicals

Malaria parasites may be killed by hydroxyl radicals *in vivo* since the effect of alloxan in diminishing the parasitaemia of *P. vinckei petteri* is reversed by desferrioxamine.[19] Alloxan reacts with tissue-derived dialuric acid to generate ROI; desferrioxamine prevents the conversion of peroxide to OH radicals by chelating the catalyst, protein-bound iron. Divicine, a derivative of alloxan, caused a similar decrease in parasitemia, and crisis forms (dead parasites within erythrocytes) and hemolysis were both noted.[20] ROI generated *in vivo* by the action of *t*-butyl hydroperoxide and iron also decreased parasitemia, and again these protective effects were reversed by desferrioxamine. Parasitized erythrocytes showed enhanced lipid peroxidation (measured by malondialdehyde concentrations) compared with uninfected cells, and this was enhanced by *t*-butyl hydroperoxide treatment and inhibited by desferrioxamine.[21]

4. Singlet Oxygen

Some of the antibody-independent inhibition of the growth of *P. falciparum* by human neutrophils may be attributed to the release of toxic singlet oxygen, since inhibition was partially abrogated by histidine and tryptophan, substances which quench singlet oxygen.[22]

B. NONOXIDATIVE KILLING

1. Tumor Necrosis Serum

Tumor necrosis serum (TNS) is generated from animals given substances such as *P. acnes* or infected with BCG or malaria parasites, to activate macrophages and then given lipopolysaccharide (LPS), which triggers the release into the serum of several cytokines, including tumor necrosis factor (TNF), and various enzymes. TNS is toxic to human and rodent malaria parasites *in vitro*[15,23,24] and it inhibits the multiplication of some, though not all, rodent parasites *in vivo*.[25] Originally its activity was thought to be due to TNF,[26] but recombinant TNF was found to be inactive (*in vitro*),[27-29] although it inhibits the multiplication of parasites *in vivo* (see Section III).

We have found that triglyceride and malondialdehyde levels in rabbit TNS are increased compared with normal rabbit serum and that, following ultracentrifugation, inhibitory activity against *P. falciparum* is found in both the lipoprotein and nonlipoprotein fractions while

TNF remains in the nonlipoprotein fraction. The serum can also be rendered nontoxic to *P. falciparum* by treatment with aerosil, a fumed silica which removes lipids. These observations are consistent with oxidized lipoproteins being at least partly responsible for the antimalarial activity in TNS,[29a] although the antiparasite effects of TNS *in vitro* cannot be inhibited by antioxidants.[30] It has been suggested that peroxidized or oxidized lipids derived by the interaction of ROI with lipids or lipoproteins may be toxic to malaria parasites. These molecules are more stable than the ROI from which they are derived, which have a half-life of fractions of seconds and whose direct action is therefore limited to the immediate vicinity of the effector cell source.

2. Eosinophil Secretory Products

The purified secretory products of eosinophils from patients with hypereosinophilia inhibited the growth of *P. falciparum in vitro*.[31] Their toxic effect was partially reversed by treatment with a monoclonal antibody against eosinophil cationic protein (ECP). ECP is highly toxic to other parasites and to mammalian cells and it causes pore formation in membranes similar to the C9 pore-forming complex of complement.[32] However, the mechanism of its toxicity for malaria parasites is unknown.

3. Polyamine Oxidase

Macrophages produce at least 80 different monokines and enzymes, in addition to ROI.[33] One macrophage-derived enzyme known to increase with activation of the cell and to be secreted extracellularly is polyamine oxidase (PAO). Human PAO catalyzes the oxidation of the ubiquitous polyamines, such as spermine, which are present in serum and in erythrocytes, to aldehydes, NH_4 and H_2O_2. Both PAO,[34,35] which is found in increased concentrations in retroplacental serum,[36] and the purified aminoaldehyde products of this reaction are toxic for *P. falciparum in vitro*.[37] Hydrogen peroxide production in this system was not considered to contribute to the inhibition seen since catalase had no effect on the killing.[35]

4. Orosomucoid/α-1 Acid Glycoprotein

Orosomucoid, an acute phase protein produced by the liver during infection, has been reported to inhibit the growth of *P. falciparum in vitro* by blocking merozoite invasion of red cells.[38] It might therefore be important in controlling parasitemia before the development of specific immunity.

5. Neutrophil-Derived Products

Polymorphonuclear cells (PMN) have been shown to kill malaria parasites by nonoxidative mechanisms since their activity was not inhibited by antioxidants or by myeloperoxidase scavengers.[39] It is possible that cationic proteins[40] or neutral proteinase[41] secreted by neutrophils may be involved. PMN from patients with chronic granulomatous disease (CGD), despite their inability to mount the respiratory burst required for production of ROI, also inhibit *P. falciparum* growth *in vitro*.[42] The nature of this toxic activity is unknown.

6. Iron and Malaria

The effect of extracellular iron on the viability of malaria parasites is a matter of much controversy.[43] On the one hand, iron deficiency has been shown to protect mice against *P. chabaudi*[44] but people with iron deficiency are still susceptible to malaria,[45] though others report an increase in parasitemia when individuals suffering from malnutrition receive iron supplements.[46]

The involvement of protein-bound iron in the generation of ROI would act to limit parasite multiplication,[19] while lactoferrin produced by PMN would chelate extracellular iron and prevent such inhibition of parasite growth.[47]

C. CELLULAR SOURCE OF TOXIC MOLECULES

1. Monocytes and Macrophages

These cells have been much studied in animal and human infections and *in vitro*. Monocytes and macrophages are widely recognized to increase in number in infection[48] and to be activated in terms of effector functions such as ROI release.[17,48-51] TNF production,[52] phagocytosis,[53,54] Fc receptors,[55] and expression of other antigens.[56]

Antibody-independent killing of rodent parasites[57,58] and of *P. falciparum*[18] *in vitro* is enhanced by cell populations enriched for monocytes or macrophages. Some of this killing may be mediated by ROI,[18] although nonoxidative killing may contribute since parasite growth is inhibited by monocytes from CGD patients whose cells are unable to mount a respiratory burst.[18] In addition, some other secretory products of activated macrophages may inhibit parasite growth.[15]

Recent work has shown that human monocytic cell lines can kill *P. falciparum in vitro* in the absence of antibody at very low effector target ratios (K. R. Jones, unpublished).

2. Neutrophils (PMN)

Malaria infection may be accompanied by a neutropenia due to PMN redistribution away from the circulation.[59] PMN may inhibit *P. falciparum* multiplication *in vitro*;[60] they appear to be activated during infection[61] and by parasite products.[62] Neutrophils can phagocytose malaria parasites,[63-65] especially opsonized free merozoites, and ROI production by PMN may be especially important since PMNs mount the most potent respiratory burst, and can be stimulated to produce superoxide which can damage red cells.[66]

3. Eosinophils

Eosinophils act as effector cells in other parasite infections, particularly by helminths.[67] Their role in protozoan infections such as malaria has not been well investigated. They disappear from peripheral blood during acute attacks but their precursor frequencies increase in the bone marrow.[68] It is possible that the mature cells are migrating to tissue sites but this appears not to have been recorded in histological studies.

In addition to the toxic effect of eosinophil secretory products already discussed, these cells have been seen to ingest rodent blood-stage parasites *in vitro* in the presence of immune serum.[64] A reduction in parasitemia in a rodent infection has been recorded when eosinophils were induced *in vivo* by other parasite antigens.[69] Eosinophils can also produce ROI.

4. Endothelial Cells

Schizonts of certain parasites, such as *P. falciparum* and *P. chabaudi*, disappear from the peripheral circulation and are sequestered in organs such as the spleen because they attach to endothelial cells in the capillaries. Endothelial cells are capable of releasing ROI and they respond to cytokines such as IL-1 and TNF so that it is possible that they might act as effector cells in some circumstances.

5. Natural Killer (NK) Cells

Parasite antigens have been shown to enhance NK activity towards tumor cells *in vitro*[70] and null cell activity is enhanced during acute human malaria.[71] However, although it has been suggested that NK cells might act as effector cells,[72,73] cross-breeding experiments with mice showed clearly that resistance to infection with *P. chabaudi*[74] and *P. vinckei*[75] is not associated with NK cell activity.

III. ROLE OF CYTOKINES IN CELL-MEDIATED IMMUNITY

Cytokines have many effects on many kinds of cells. They may accelerate and enhance immune responses, but they may also suppress them, for example, by down-regulating

TABLE 1

Cytokines Involved in Cell-Mediated Responses, the Principal Cells from Which They are Derived, and Some Stimuli Inducing Their Secretion

Cytokine	Cell of origin	Inducing stimulus
IL-1	Macrophage and various	Yeasts, bacteria, viruses immune complexes
TNF	Macrophage lymphocyte	Yeasts, bacteria, viruses,
LT		specific antigen
IL-2	T cell	Specific antigen
IL-3 (CSF a)	T cell	Specific antigen
IL-4 (BSF-1)	T cell	Specific antigen
IL-5 (EDF)	T cell	Specific antigen
IL-6 (BSF-2, β_2 IFN)	T cell	Specific antigen
G-M CSF	T cell, macrophage, endothelial cell, fibroblast	Specific antigen bacteria, other cytokines
M-CSF	Macrophage	Bacteria, other cytokines
IFN-α	Leukocytes, various	Viruses, poly I: poly C
IFN-β	Fibroblast, various	Viruses, poly I: poly C
IFN-γ	T cell	Specific antigen
	B cell, macrophage	Various

receptors or by inhibiting cellular proliferation. Many of them cause increased secretion of other cytokines, some even acting in autocrine fashion to induce further secretion of the same cytokine itself, and since many also act synergistically with others substantial amplification of their effects can occur. Cytokines can be grouped broadly into those derived from lymphocytes, particularly T cells, mainly in response to specific antigens and those secreted by macrophages, usually in response to microbial stimuli (Table 1). Some, however, may be secreted by both and some by fibroblasts and endothelial cells.

None of these molecules is known to be directly cytotoxic to any forms of the malarial parasite *in vitro*, in the absence of effector cells. Several can cause a reduction in parasite numbers indirectly, however, by their effects on hepatocytes or monocytes in culture and two of them, TNF and IFN-γ, have been shown to induce protective effects *in vivo* against both the liver and blood stages of the parasite (see Section III).

Since the advent of recombinant cytokines, our knowledge of their involvement in inflammation and infection has increased dramatically. Apart from acting as mediators of inflammation in general, they may play a key role by enhancing the ability of various effector cells to kill the parasites; indeed, some of the protective effects of vaccination may be mediated by the accelerated production of certain cytokines. A number, alone or in combination, enhances various functions of different cells with consequences that are potentially parasiticidal; some examples are listed in Table 2. It is likely that all increase in amount in malaria, at least in the early stages of infection before immunosuppression intervenes, but few systematic studies have been reported (see Section III.B.1). While small increases for short periods may be beneficial, excess secretion of some cytokines during the acute phase of infection may be responsible for some of the pathology of the disease (see Chapter 6).

A. ANTIGENIC STIMULI

Little is known about the nature of malarial antigens capable of stimulating secretion of the various cytokines and it is possible that epitopes required for CMI may differ from those required for antibody production. For example, MHC Class II restricted epitopes on the circumsporozoite protein which are recognized by CD4+ cells did not prime T cells for IFN-γ-mediated protection.[113] Similarly, antigens reacting with lymphocytes may differ from those that bind to receptors on macrophages and stimulate the release of their cytokines.

TABLE 2
Examples of Enhancement by Purified or Recombinant Cytokines of Potentially Parasiticidal Mechanisms of Effector Cells

Effector Cell	Function enhanced	Cytokine	Ref.
Macrophages and monocytes	Phagocytosis Fc receptors	GM-CSF	76
		IL-4	77
		M-CSF	78
		IFN-γ	79,80
	ROI	IL-1	81
		TNFs	82,83
		IFN-γ	84 — 86
		GM-CSF	83
		M-CSF	83,87
	Killing of tumor cells	IL-1	88
		TNFs	82
		IL-4	77
		GM-CSF	89
		IFN-γ	90
	Resistance to *Trypanosoma cruzi*	TNFs	91
		GM-CSF	83
		IFN-γ	83
	Killing of *Toxoplasma gondii*	IFN-γ	84
	Leishmania donovani	GM-CSF	92
	Schistosoma mansoni	TNFs	82
		IFN-γ	93
	Chemotaxis	TNF	94
Neutrophil	Phagocytosis	TNFs	95
		GM-CSF	96
		IL-1	81
	ROI	TNFs	95 — 98
		GM-CSF	96,99
	ADCC	TNFs	100
		GM-CSF	76
	Killing of *Candida albicans*	TNFs	101
		IFN-γ	101
	Trypanosoma cruzi	CM-CSF	102
	Chemotaxis	IL-1	103
		TNF	94
Eosinophil	Phagocytosis	IL-3	104
	Degranulation	IL-1	105
	ROI	IL-3	104
		GM-CSF	96
	ADCC	IL-3	104
		IL-5	106
	Killing of *S. mansoni*	TNFs	107 — 109
		GM-CSF	106,110
Endothelial cell	ROI	IL-1	111
		IFN-γ	111
NK cells	Killing of tumor cells	TNF	112
		IFN-γ	112

Note: Some require the presence of another cytokine and some effects may be mediated through induction of secretion of another cytokine.

Some antigens appear to be mitogenic and their action does not require recognition by specific T cells but may nevertheless be accompanied by the release of some cytokines. Sporozoites of *P. berghei* stimulate proliferation of normal murine spleen cells *in vitro*[114] and the blood stage of *P. falciparum*[115] and some substances released in culture[116-118] can stimulate proliferation of normal human lymphocytes, though soluble antigens did not do so when affinity purified.[119] On the other hand, failure of lymphocytes to respond during infection would be accompanied by a failure of production of lymphokines secreted by T cells. Thus, Ho and co-workers[120] observed diminished proliferation, in response to malarial antigens, of T cells from patients infected with *P. falciparum* which returned to normal as parasites were cleared; suppression correlated with the initial parasite count and the severity of clinical illness, and it was more prolonged in those with serious disease.

Parasites can also activate macrophages *in vivo* independently of T cells, in that the degree of activation of macrophages from the livers of nude mice infected with *P. berghei*, measured by the production of plasminogen activator, and the increase in macrophage yield and liver weight did not differ from controls.[121] The same was true for the yield of TNF in the serum of infected nude mice given endotoxin, another marker of macrophage activation (unpublished work). Furthermore, soluble antigens from parasite cultures induce the release of some cytokines *in vivo* and also from activated macrophages *in vitro* (see below).

B. T CELL FACTORS

Although it is well recognized that immunity to both sporozoite and blood stage infections depends upon the presence of T cells, there is little evidence to suggest that T cells kill parasites directly. Their presence may be essential because in response to malarial antigens they secrete mediators which are required for the expansion and activation of other cell populations, in addition to IL-2, the molecule required for the expansion of the T cell population itself. Factors such as IL-3, IL-5, and the colony-stimulating factors, for instance, act as growth factors for myeloid cells and IL-4, IL-6, and IFN-γ can cause their activation. IL-3 is released soon after antigenic stimulation[122] and could be one of the essential mediators of CMI, but no studies have yet been published on its involvement in malaria. GM-CSF enhances ROI production from macrophages faster than does IFN-γ[83] and it is likely to be active earlier since it is also secreted by macrophages, endothelial cells, and fibroblasts and can therefore appear before significant T cell proliferation has occurred. However, the only published studies on T cell factors are those concerned with IL-2 and IFN-γ, both of which, it is interesting to note, can enhance ROI production by monocytes.[84,123]

As the degree of protection seems to depend upon the kinetics of an early immune reponse,[124] which must outpace the rate of parasite growth,[125] immune T cells (memory cells) may be protective because they respond faster to infection specifically by releasing mediators like IFN-γ which enhance nonspecific effector reactions.

The effectiveness of a protective response depends upon quantitative changes in effector cell populations, many of which are T cell dependent, as well as upon qualitative ones. For example, Roberts and Weidanz[126] observed that splenomegaly and enhanced phagocytosis in mice infected with *P. yoelii* were thymus dependent. Myeloid cells in the blood,[48,127] spleen,[48,128] and liver[129] increase in number during infection, as well as in activity, and this increase is enhanced in vaccinated animals.[48,129] Indeed, compared with controls, macrophages in both the spleens and livers of mice vaccinated against the blood stage of a lethal variant of *P. yoelii* were activated more rapidly after challenge, both in terms of their capacity to give an oxidative burst and their cytotoxic activity against tumor cells, and a striking increase occurred in the yield of infiltrating cells obtained from the liver.[130] Immunity seems to be associated with a blood monocytosis,[48] which in the case of *P. yoelii* blood-stage infections at least appears to be related to factors produced by sensitized Lyt-1 T cells,[131] and blood monocytes are recruited to the spleen.[128]

1. Interleukin-2

To date, the results of investigations on the production of this lymphokine in malaria suggest that the failure of host effector cells to control parasite multiplication, and the well-known immunosuppression associated with malaria, may be related to a lack of sufficient IL-2 to support proliferation of the necessary T cells. Thus, although *P. falciparum* antigens purified by immunoadsorption from culture supernatants induced proliferation of T helper cells from immune individuals specifically, IL-2 production was observed only from those giving a strong response.[119] T cells taken from Colombian patients during the acute stage of infection with *P. falciparum* gave only a weak and short-lived proliferative response to specific antigen, and this was enhanced and prolonged by addition of IL-2;[132] similarly, acutely infected patients from Thailand were defective in IL-2 production when exposed to malarial antigens.[133]

Studies on the ability of T cells from the spleens of mice infected with *P yoelii* and *P. berghei* to secrete IL-2 in response to mitogenic stimulation suggest that infection causes a block in its production.[134] In both infections, the amount of IL-2 released diminished progressively from the 3rd day until there was virtually none detectable as mice infected with *P. berghei* began to die. In the nonlethal infection, however, IL-2 production began to increase as the parasitemia was controlled to reach a level significantly above normal as parasites disappeared from the blood. The diminished response to mitogen was not due to the presence of suppressor cells in the spleen, but was apparently due to a failure of the cells to secrete IL-2 — they responded normally if cultured with a preparation containing exogenous IL-2.

It has also been shown that during infection a substance appears in the serum which both blocks secretion of IL-2 and neutralizes preexistent IL-2, and which would therefore have a profound effect on the regulation of CMI. This inhibitor was detected in mice infected with *P. yoelii* or *P. berghei*.,[135] it increased in amount soon after infection to reach one peak after 3 d and a second after 10 d, and it returned to normal levels in mice infected with *P. yoelii* as they recovered. Its production was shown to be T cell dependent. It resolved to a single peak following isoelectric focusing and ion exchange chromatography and might possibly represent the serum analogue of the IL-2 receptor.[136]

2. Interferons

Although not all IFNs are products of T cells, they are included together in this section because they may often, though not always, have similar actions but with different potency. Interferons can influence cell behavior in a number of ways. By regulating the expression of specific genes they can, for example, modulate the expression of cell surface receptors, affect cell differentiation, inhibit cell growth, and proliferation, activate effector cells, including macrophages and NK cells, and inhibit virus replication. Furthermore, it has been observed that more IFN-γ is induced during the development of a CMI response to antigen than an antibody response.[137]

Interferon-α and-γ have been detected in the serum of patients shortly after an acute attack of *P. falciparum* malaria,[138] and in children the presence of IFN, predominantly IFN-α, correlated with the degree of parasitaemia.[71] The antigens involved have not been identified or related to the developmental stage of the parasite; however, polyclonal T cell proliferation induced by sporozoites of *P. berghei* was accompanied by the production of IFN-γ[114] and IFN activity was detected in the serum of mice within hours of infection with blood stage *P. berghei*,[139] although the type of IFN is unknown. *In vitro*, free *P. falciparum* parasites, but not intraerythrocytic forms, were found to induce production of IFN-α from human peripheral blood "null" cells.[140]

a. *Effect of IFNs*

Known inducers of IFN-α and IFN-β reduced the severity of *P. berghei* infection initiated by sporozoites but not by blood forms.[141] Similarly, injection of serum containing virus-induced IFN diminished the parasitemia of mice infected with *P. berghei* sporozoites, again apparently by its action on the preerythrocytic stage.[142] Furthermore, with blood stage *P. berghei*, administration of antibody against IFN-β to infected mice accelerated the development of parasitemia only slightly.[143] Various nonspecific macrophage-activating agents such as BCG, *C. parvum*, and endotoxin increase resistance to malaria and may act by enhancing the production of cytokines, including IFNs (reviewed by Playfair[144]), and again it is interesting that IFN inducers and *C. parvum* are more effective against sporozoite-induced *P. berghei* infections, especially at the liver stage, than against the blood stage. Experiments with recombinant cytokines support this interpretation of the mode of action of agents inducing resistance to malaria and exoerythrocytic forms seem to be more susceptible than blood stage parasites.

b. Exoerythrocytic Forms

Administration of human IFN-γ to chimpanzees infected with sporozoites of *P. vivax* diminished parasitemia[145] and higher doses totally prevented the appearance of parasites in the blood of rhesus monkeys infected with sporozoites of *P. cynomolgi*.[146] In rats and mice infected with sporozoites of *P. berghei*, IFN was shown to inhibit the development of exoerythrocytic forms (EEF) in the liver;[147] this was not due to a direct effect on the sporozoites, which are cleared from the circulation in 30 min and which were injected $2^1/_2$ h before IFN-γ.[113] In theory, IFN-γ *in vivo* might act via effector cells but there is no evidence that inflammatory cells are drawn to the liver in that stage of the infection, and it appears to have a direct effect on the liver cells themselves. Thus, EEF failed to develop *in vitro* in a hepatoma cell line treated with the cytokine before infection with *P. berghei* sporozoites.[147] The cells express surface receptors for IFN-γ[148] and most cells were rendered resistant to sporozoites by pretreatment.[149]

Similarly, IFN-γ inhibited the development of schizonts of *P. falciparum* in cultures of human hepatocytes, when added after addition of sporozoites, and even caused the disappearance of schizonts when added after 4 d of culture.[150] Recombinant IFN-α and IFN-β also inhibited the development of schizonts but 1000-fold more was required. *In vivo* it appears that CD8+ T cells may be involved in IFN-γ-mediated protection of immunized mice, since administration of a monoclonal antibody specific for these T cells, or of another antibody which neutralized IFN-γ, abolished their immunity to challenge by *P. berghei* sporozoites.[113] Immunity transferred by T cells was reversed by antibody against IFN-γ.

c. Blood-Stage Parasites

In contrast to the direct effect of IFNs on the liver cell inhabited by the parasite, neither IFN-α[151] nor IFN-γ[152] inhibited the multiplication of *P. falciparum* within erythrocytes *in vitro*. However, the ability of human monocytes to inhibit the multiplication of *P. falciparum* within erythrocytes was enhanced by IFN-γ.[18] Crisis forms were seen and enhancement was associated with an increased capacity of the cells to secrete H_2O_2 when stimulated with PMA. Similarly, mouse peritoneal macrophages that had been activated by a lymphokine preparation which would have contained IFN-γ showed an enhanced ability to kill *P. yoelii* through the release of toxic soluble factors.[153] *In vivo*, administration of high doses of IFN-γ to mice infected with blood-stage parasites of *P. berghei* did not significantly delay the time course of the parasitemia,[145] but both specific and nonspecific immunity against the blood stages of this parasite are notoriously difficult to induce.

C. MACROPHAGE FACTORS

The two main cytokines secreted by macrophages, IL-1 and TNF, have a similar range of activities in many respects,[154] although they are not always equally potent in all their capacities. Macrophages, when activated, secrete more IL-1 and TNF[155] and there is plenty of evidence that they are activated in malaria, in rodents,[25,49,50,52,156] and in man.[55] Macrophages can be activated by a number of T cell products and by other cytokines, including IL-3, IL-4, GM-CSF, lymphotoxin, and IFN-γ, and possibly by products of parasites themselves. IL-1 was first identified as the primary mediator of T cell proliferation but it now seems that TNF can also stimulate the growth of T cell colonies[157] and exert many stimulatory effects on activated T cells, including enhancing their proliferative response to IL-2.[158,159] Workers in our field have generally been interested in one or other of these molecules but not both, and until recently have tended to regard IL-1 as a lymphocyte-activating factor and TNF as a potentially cytotoxic molecule, so no precise comparisons of their production or activities in malaria have been made. It should be kept in mind, however, that they may be released from the same type of cell, sometimes in the same conditions, that each can enhance the production of the other,[16] so that they may act synergistically,[161,162] and that T cell proliferation apart they may often be responsible for many of the same effects, indirectly if not directly. It has also been reported, however, that IL-1 can induce cells to become resistant to TNF.[163] Lymphotoxin appears to have a similar range of biological activities to TNF and, as it binds to the same receptor, it is assumed here that the effects of any lymphotoxin produced are likely to resemble those of TNF.

Apart from the fact that IL-1 production is fundamental to the development of any immune response, since T cell proliferation depends upon it, IL-1 and TNF may help the host to eliminate parasites by influencing cell migration, since they can act synergistically to induce neutrophil migration,[164] and by enhancing the activity of effector cells. Both, for example, enhance phagocytosis and increase the production of ROI by neutrophils,[81,98] IL-1 induces the release of ROI from endothelial cells (as does IFN-γ),[111] and TNF has been shown to cause degranulation of neutrophils[95] to enhance their ADCC activity[100] and to activate eosinophils.[109] Some of these effects may be further enhanced by the presence of IFN-γ[98], which is known also to enhance the production of IL-1, TNF, and lymphotoxin.

1. Interleukin-1 and Its Effects

Studies on the production of IL-1 *in vivo* in response to the conventionally used stimulant bacterial LPS indicate that cells become capable of producing more of this cytokine during infection. Progressively more IL-1 was detected in the serum of mice 2 h after administration of LPS at different times during the course of infection with *P. vinckei-petteri*, and the concentration reached a plateau while the parasitemia was still rising; peritoneal macrophages taken at one time point when the parasitemia was high secreted more IL-1 in response to a low dose of LPS than cells from uninfected mice.[165] Spleen macrophages taken from mice throughout the course of infection of *P. yoelii* and stimulated with LPS yielded more IL-1 than controls, especially on day 10 and as the mice recovered, whereas those from mice infected with *P. berghei* did not.[134] Peritoneal cells, on the other hand, responded in both cases with an early peak a few days after infection; levels subsequently dropped, to rise again during the latter part of the *P. yoelii* infection. Spira and co-workers[166] found no increase in the yield of IL-1 induced by LPS from spleen macrophages taken from rats infected with *P. berghei* at a single time point, day 13.

Macrophages secrete IL-1 spontaneously during infection, presumably in response to parasite products. Thus, *P. berghei* antigens were shown to induce secretion of IL-1 from spleen macrophages from normal mice[167] and, a few days after infection with *P. berghei*, mouse spleen and peritoneal macrophages were found to release IL-1 spontaneously.[134,167] Spleen macrophages[167] and peritoneal cells[134] from mice infected with *P. yoelii* also showed enhanced secretion early on, and levels were greatly above normal as the mice recovered. However, supernatants from spleen macrophages taken during the middle of either infection inhibited T cell proliferation in response to mitogens as well as to antigen preparations of

P. berghei and *P. yoelii*.[167] Such suppressive activity might perhaps explain the lack of enhancement of IL-1 production observed for spleen macrophages from mice[134] and rats[166] infected with *P. berghei*.

There are no reports of the effects of administration of IL-1 to infected hosts, but it would seem probable that like IFN-γ and TNF (see below) it would be especially effective against the liver stage since treatment of human or rat hepatocytes before infection with sporozoites of *P. falciparum* or *P. yoelii*, respectively, have recently been shown to inhibit the development of schizonts.[150] Unlike IFN-γ, however, treatment after inoculation of sporozoites had no effect.

2. TNF and Its Effects

There are few reports of the detection in the circulation of free TNF spontaneously produced during infection; this is not surprising considering the short half-life of this cytokine *in vivo* and the fact that excess production is likely to be associated with severe and prolonged morbidity. Scuderi and co-workers,[168] using an enzyme-linked immunosorbent assay, found TNF in the serum of seven out of ten patients with malaria (clinical details not given), and others have detected it in the serum of patients infected with *P. falciparum* (Hommel, personal communication). In Chapter 6, however, Clark and Chaudhri detected TNF in the serum of mice infected with *P. vinckei*, but only towards the end of the (lethal) infection, and similarly trace amounts were found in the serum of CBA mice infected with *P. berghei* ANKA when they were dying prematurely with neurological symptoms, though not in those dying later with a high parasitemia.[169] Using a sensitive cytotoxicity assay, we ourselves have never found significant or reproducible amounts in mouse serum at any time during the course of lethal (*P. berghei*, *P. yoelii*) or nonlethal (*P. yoelii*, *P. chabaudi*) infections[25] (and unpublished work).

It is clear, however, that one of the early host reponses to infection is an enhanced capacity to produce TNF. There are as yet no reports of TNF production in response to sporozoite-induced infections but several concerning blood-stage infections. The appearance of LPS-induced TNF in the serum of mice infected with *P. yoelii* or *P. berghei* ran parallel to the course of parasitemia[25] and sera from mice infected with *P. vinckei-petteri* and given LPS in midinfection contained TNF.[170] Investigation of the capacity of mouse spleen macrophages to kill L 929 tumor cells, known to be mediated by TNF, increased dramatically soon after infection with *P. yoelii* or *P. chabaudi*, though to a lesser degree with a lethal form of *P. yoelii* or with *P. berghei*, and the development of this cytotoxicity was faster and often greater in vaccinated mice.[52,130] Liver macrophages are also able to secrete more TNF in malaria, as shown by the increased TNF-mediated cytotoxicity of such cells isolated from mice infected with lethal[130] or nonlethal *P. yoelii* (quoted by Dockrell[50]).

Activated macrophages do not release TNF spontaneously but only in response to stimulation (or "triggering"). Macrophages from mice with malaria can secrete TNF in response to exogenous stimulants, such as that provided by the target tumor cell in a cytotoxicity assay or by bacterial endotoxin, and recently we have shown that parasites and some of their soluble antigens can themselves induce its release.[171a,171b] Mouse erythrocytes infected with nonlethal or lethal *P. yoelii* or with *P. berghei*, in the presence of polymyxin B to exclude the effects of any contaminating endotoxin, induced activated mouse peritoneal macrophages to produce as much TNF as did microgram amounts of LPS; they were still active after fixation with formalin. Parasites incubated overnight released soluble products which triggered secretion of TNF and which withstood boiling. Their ability to induce TNF release was inhibited by specific antiserum. Human monocytes, activated by pretreatment with IFN-γ and indomethacin, were shown to secrete TNF in response to these boiled antigen preparations.[171c] Formalin-fixed erythrocytes infected with *P. falciparum* also stimulated secretion of TNF, as did concentrated medium from continuous cultures. Both human monocytes and mouse macrophages released TNF in response to incubation with a soluble antigen of *P. falciparum* purified by affinity chromatography from culture supernatants (unpublished work in collaboration with Jakobsen et al.). This antigen, and some others, bind to Limulus lysate and to polymyxin B,[172] perhaps settling the old controversy about the presence of endotoxin-like activity in the serum of patients infected with *P. falciparum*. Heat-stable antigens of *P. berghei* and of *P. yoelii* also act like LPS *in vivo* to release TNF, since they kill mice given D-galactosamine,[171b] which sensitizes them to the toxic effects of TNF.[173] It thus seems highly likely that during infection TNF is produced by activated macrophages in response to stimulation by soluble antigens released by the parasite; their nature has not yet been determined but they resemble S antigens in their ability to withstand boiling.

a. Exoerythrocytic Forms

As with IFN-γ, the effect of TNF against the early stages of the parasite is very striking. Administration of recombinant human TNF by osmotic pumps to mice subsequently infected with *P. berghei* sporozoites completely inhibited the development of the parasite (Schofield & Ferreira, quoted by Nussenzweig[174]). TNF probably acted by exerting a direct effect on the liver cells since it diminished the amount of parasite DNA detectable in the livers of rats given TNF and then challenged with *P. berghei* sporozoites and it also prevented their development into EEF in a hepatoma cell line.[174]

b. Blood-Stage Parasites

Although, as mentioned before, it seemed that TNF was toxic to parasites *in vitro* since the parasite-killing component of TNS copurified with tumor cell cytotoxicity,[26] it is now clear that TNF itself does not kill parasites directly[27-29] even in the presence of IFN-γ, so that the antiparasitic effects *in vivo* must be mediated via an as yet unidentified TNF responsive cell or cells. The multiplication of *P. yoelii* in mice is inhibited by rTNF administered at the start of infection and the survival of mice infected with a lethal variant was significantly prolonged.[27] Administered in osmotic pumps, it greatly suppressed the parasitemia of *P. chabaudi*, induced the appearance of crisis forms,[175] and it significantly decreased the parasitemia of mice infected with *P. berghei* (Schofield and Ferreira quoted by Nussenzweig[174]). The latter effect was mediated mainly by the spleen in that only small differences were observed in splenectomized mice, indicating that TNF-responsive cells reside there. How they control parasitemia is unknown.

IV. T CELLS AND THE EFFECT OF VACCINATION

The involvement of T cells in malaria is complex. To begin with, T cell numbers are frequently reduced during infection in both humans and experimental animals.[144] In some cases there may be actual loss of T cells due to autoantibodies,[176] but there may also be shifts of T cells out of the blood into organs such as the liver and spleen.[177] T cell responses are also frequently suppressed during blood-stage infection, examples being contact sensitivity,[178] T-dependent antibody production,[179] and proliferative responses to mitogens *in vitro*.[180]

Conversely, increased T cell function often correlates with useful immunity. For example, mice vaccinated against blood-stage *P. yoelii*, which gave good delayed hypersensitivity responses to parasite antigen injected into the ear, were also likely to be well protected against a subsequent infection.[1] In humans, proliferative responses to parasite antigens are usually greater in immune than in nonimmune individuals[181,182] and the same is true, though not so consistently for IFN-γ production *in vitro*.[132] A problem with such experiments is selecting the appropriate antigen; for example, it may be that responses to the intact parasitized red cell and to soluble released antigens might have quite different effects on the infection and therefore different prognostic significance.

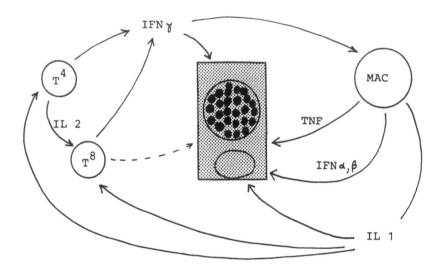

FIGURE 1. Cell-mediated immunity and the liver stage. Left: antigen-specific and right: nonantigen-specific. T⁴, T⁸: CD 4+, CD 8+ T lymphocytes. MAC: macrophage; IFN: interferon: TNF: tumor necrosis factor; and IL: interleukin.

It has not always proved easy to identify the precise role of T cells in immunity. Phenotypic analysis based on the effects of monoclonal anti-T cell subset antibodies in mouse experiments suggests that in the liver stage it is the CD8+ cell (cytotoxic/suppressor) which is most important,[113] while the CD4+ (helper) cells are required for immunity to the asexual blood stage[183] though in one model there was also evidence for a transient role for CD8+ cells in the late stage of blood-stage infection.[131] Unfortunately, there is at present no reliable method of separately depleting those CD4+ cells that help B cells from those that activate macrophages. In experiments with *P. yoelii* antigens separated by isoelectric focusing, we showed that T help for antibody and the induction of DTH were always found together (and usually correlated well with protection against challenge[184]). Of ten human T cell clones from an immune patient, two CD4+ and one CD8+ clone responded to a semipurified 50-kDa antigen, while all six CD4+ and all four CD8+ clones responded to parasitized red cells by both proliferation and IFN release.[185] Clearly the possible combinations of T cell and antigen for this type of experiment must run into thousands, but it should in principle be feasible to do large-scale testing using the recently described method of antigen blotting.[186] Meanwhile one can only speculate on the way in which T cells may work.

V. SOME SPECULATIONS AND A HYPOTHESIS

A. LIVER STAGE

Here it is clear that T cells have at least two roles: CD4+ cells help to induce anti-sporozoite antibody while CD8+ cells presumably protect the infected liver cell, either by releasing IFN-γ or by direct cytotoxicity. The IFN-γ might act by directly activating an anti-parasitic function of the liver cell, somewhat in the manner of its antiviral activity, or by enhancing the response to low levels of other cytokines. The implication for vaccination would be that quite complex antigens or mixtures of antigens might be required for maximal immunity against this stage (see Figure 1).

B. ASEXUAL BLOOD STAGE

Because of its distribution, mobility, and repetitive life cycle, this stage is potentially exposed to virtually all the cellular and humoral components of the immune system so that any mechanism with demonstrable antiparasitic activity needs to be considered. This would

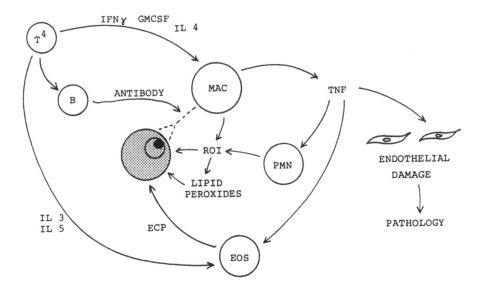

FIGURE 2. Cell-mediated immune responses and the asexual blood stage. PMN: polymorphonuclear neutrophil; EOS: eosinophil; ROI: reactive oxygen intermediates; and ECP: eosinophil cationic proteins. B: B lymphocyte; T⁴: CD4⁺ lymphocyte; GMCSF: granulocyte-momocyte colony stimulating factor.

include the ROIs (perhaps acting through the much more stable oxidized lipids) and the eosinophil cationic proteins (see Section II.B.1. and 2) plus any cell or cytokine capable of inducing these.

Figure 2 is an attempt to bring all these elements together in a single scheme, which is no doubt both oversimplified and incomplete. It does emphasize however, how vulnerable the blood-stage parasite is to attack, and how important it is to understand how it avoids destruction. Regarding CMI, several mechanisms come to mind. Antigenic variation might restrict the effectiveness of T cells against repeated attacks just as it frustrates antibody; it is obviously important to assess the degree of this using cloned T cells and antigens. Parasites might be expected to sequester in particular sites where cell-mediated responses are ineffective or are suppressed because they are dangerous. The attachment of *P. falciparum* to brain endothelium may be an example of this, and cerebral malaria may represent the result of insufficiently suppressed cell-mediated activity; in mice TNF appears to be the principal mediator involved but there is little evidence that this is so in the human disease. Nonetheless, the association between excessive levels of TNF and other cytokines and severe pathological processes noted earlier (see Chapter 6) suggests to us that the single most important feature of all (the cell-mediated) pathways considered in this chapter is whether they are focused sharply onto the parasite, when they are likely to contribute significantly to protective immunity, or disseminated in a body-wide fashion, in which case the predominant effect will be pathological. Our working hypothesis is that it is the role of the adaptive immune system — both T cells and antibody — to ensure that this focusing occurs.

ACKNOWLEDGMENTS

The work reported here from our department was supported by grants from the Medical Research Council of Great Britain and the Wellcome Trust.

REFERENCES

1. **Cottrell, B. J., Playfair, J. H. L., and de Souza, J. B.,** Cell-mediated immunity in mice vaccinated against malaria, *Clin. Exp. Immunol.,* 34, 147, 1978.
2. **Taliaferro, W. H. and Taliaferro, L. G.,** The effect of immunity on the asexual reproduction of *Plasmodium brasilianum, J. Infect. Dis.,* 75, 1, 1944.
3. **Cohen, S., McGregor, I. A., and Carrington, S. C.,** Gamma globulin and acquired immunity to human malaria, *Nature,* 192, 733, 1961.,
4. **Clark, I. A., Allison, A. C., and Cox, F. E.,** Protection of mice against *Babesia* and *Plasmodium* with BCG, *Nature,* 259, 309, 1976.
5. **Clark, I. A., Cox, F. E. G., and Allison, A. C.,** Protection of mice against *Babesia* spp. with killed *Corynebacterium parvum, Parasitology,* 74, 9, 1977.
6. **Grun, J. L. and Weidanz, W. P.,** Immunity to *Plasmodium chabaudi adami* in the B cell deficient mouse, *Nature,* 290, 143, 1981.
7. **Grun, J. L. and Weidanz, W. P.,** Antibody-independent immunity to reinfection malaria in B-cell-deficient mice, *Infect. Immun.,* 41, 1197, 1983.
8. **Friedman, M. J.,** Oxidant damage mediates variant red cell resistance to malaria, *Nature,* 280, 245, 1979.
9. **Etkin, N. L. and Eaton, J. W.,** Malaria-induced erythrocyte oxidant sensitivity, in *Erythrocyte Structure and Function,* Vol. 1, Brewer, G. J., Ed., Alan R. Liss, New York, 1975, 219.
10. **Fairfield, A. S., Meshnick, S. R., and Eaton, J. W.,** Malaria parasites adopt host cell superoxide dismutase, *Science,* 221, 764, 1983.
11. **Stocker, R., Hunt, N. H., Buffinton, G. D., Weidemann, M. J., Lewis-Hughes, P. H., and Clark, I. A.,** Oxidative stress and protective mechanisms in erythrocytes in relation to *Plasmodium vinckei* load, *Proc. Natl. Acad. Sci. U.S.A.,* 82, 548, 1985.
12. **Trager, W. and Jensen, J. B.,** Human malaria parasites in continuous culture, *Science,* 193, 673, 1976.
13. **Allison, A. C. and Eugui, E. M.,** The role of cell-mediated immune response in resistance to malaria, with special reference to oxidant stress, *Ann. Rev. Immunol.,* 1, 361, 1983.
14. **Weed, R. I.,** Effects of primaquine on the red blood cell membrane. II. K^+ permeability in glucose-6-phosphate dehydrogenase deficient erythrocytes, *J. Clin. Invest.,* 40, 140, 1961.
15. **Wozencraft, A. O., Dockrell, H. M., Taverne, J., Targett, G. A. T., and Playfair, J. H. L.,** Killing of human malaria parasites by macrophage secretory products, *Infect. Immun.,* 43, 664, 1984.
16. **Dockrell, H. M. and Playfair, J. H. L.,** Killing of *Plasmodium yoelii* by enzyme-induced products of the oxidative burst, *Infect. Immun.,* 43, 451, 1984.
17. **Wozencraft, A. O., Croft, S. L., and Sayers, G.,** Oxygen radical release by adherent cell populations during the initial stages of a lethal rodent malarial infection, *Immunology,* 56, 523, 1985.
18. **Ockenhouse, C. F., Schulman, S., and Shear, H. L.,** Induction of crisis forms in the human malaria parasite *Plasmodium falciparum* by γ-interferon-activated, monocyte-derived macrophages, *J. Immunol.,* 133, 1601, 1984.
19. **Clark, I. A. and Hunt, N. H.,** Evidence for reactive oxygen intermediates causing hemolysis and parasite death in malaria, *Infect. Immun.,* 39, 1, 1983.
20. **Clark, I. A., Cowden, W. B., Hunt, N. H., Maxwell, L. E., and Mackie, E. J.,** Activity of divicine in *Plasmodium vinckei*-infected mice has implications for treatment of favism and epidemiology of G-6-PD deficiency, *Br. J. Haematol.,* 57, 479, 1984.
21. **Clark, I. A., Hunt, N. H., Cowden, W. B., Maxwell, L. E., and Mackie, E. J.,** Radical-mediated damage to parasites and erythrocytes in *Plasmodium vinckei* infected mice after injection of t-butyl hydroperoxide, *Clin. Exp. Immunol.,* 56, 524, 1984.
22. **Nnalue, N. A. and Friedman, M. J.,** Evidence for a neutrophil-mediated protective response in malaria, *Parasite Immunol.,* 10, 47, 1988.
23. **Taverne, J., Dockrell, H. M., and Playfair, J. H. L.,** Endotoxin-induced serum factor kills malarial parasites *in vitro, Infect. Immun.,* 33, 83, 1981.
24. **Haidaris, C. G., Haynes, J. D., Meltzer, M. S., and Allison, A. C.,** Serum containing tumor necrosis factor is cytotoxic for the human malaria parasite *Plasmodium falciparum, Infect. Immun.,* 42, 385, 1983.
25. **Taverne, J. Depledge, P., and Playfair, J. H. L.,** Differential sensitivity *in vivo* of lethal and non-lethal malarial parasites to endotoxin-induced serum factor, *Infect. Immun.,* 37, 927, 1982.
26. **Taverne, J., Matthews, N., Depledge, P., and Playfair, J. H. L.,** Malarial parasites and tumour cells are killed by the same component of tumour necrosis serum, *Clin. Exp. Immunol.,* 57, 293, 1984.
27. **Taverne, J., Tavernier, J., Fiers, W., and Playfair, J. H. L.,** Recombinant tumour necrosis factor inhibits malaria parasites *in vivo* but not *in vitro, Clin. Exp. Immunol.,* 67, 1, 1987.
28. **Jensen, J. B., Vande Waa, J. A., and Karadsheh, A. J.,** Tumor necrosis factor does not induce *Plasmodium falciparum* crisis forms, *Infect. Immun.,* 55, 1722, 1987.

29. **Hviid, L., Reimert, C. M., Theander, T. H., Jepsen, S., and Bendtzen, K.**, Recombinant human tumour necrosis factor is not inhibitory to *Plasmodium falciparum in vitro, Trans. R. Soc. Trop. Med. Hyg.*, 82, 48, 1988.

29a. **Rockett, K. A., Targett, G. A. T., and Playfair, J. H. L.**, Killing of blood-stage *Plasmodium falciparum* by lipid peroxides from tumor necrosis serum, *Infect. Immun.*, 56, 3180, 1988.

30. **Geary, T. G., Boland, M. T., and Jensen, J. B.**, Antioxidants do not prevent the *in vitro* induction of *Plasmodium falciparum* crisis forms by human malaria-immune, TB or rabbit TNF serum, *Am. J. Trop. Med. Hyg.*, 35, 704, 1986.

31. **Waters, L. S., Taverne, J., Tai, P.-C., Spry, C. J. F., Targett, G. A. T., and Playfair, J. H. L.**, Killing of *Plasmodium falciparum* by eosinophil secretory products, *Infect, Immun.*, 55, 877, 1987.

32. **Ding-E. Young, J., Peterson, C. G. B., Venge, P., and Cohn, Z. A.**, Mechanism of membrane damage mediated by human eosinophil cationic protein, *Nature,* 321, 613, 1986.

33. **Nathan, C. F.**, Secretory products of macrophages, *J. Clin. Invest.*, 79, 319, 1987.

34. **Ferrante, A., Rzepczyk, C. M., and Allison, A. C.**, Polyamine oxidase mediates intraerythrocytic death of *Plasmodium falciparum, Trans. R. Soc. Trop. Med. Hyg.*, 77, 789, 1983.

35. **Rzepczyk, C. M., Saul, A. J., and Ferrante, A.**, Polyamine oxidase-mediated intraerythrocytic killing of *Plasmodium falciparum*: evidence against the role of reactive oxygen metabolites, *Infect. Immun.*, 43, 238, 1984.

36. **Egan, J. E., Haynes, J. D., Brown, N. D., and Eisemann, C. S.**, Polyamine oxidase in human retro-placental serum inhibits the growth of *Plasmodium falciparum, Am. J. Trop. Med. Hyg.*, 35, 890, 1986.

37. **Morgan, D. M. L., Bachrach, U., Assaraf, Y. G., Harari, E., and Golenser, J.**, The effect of purified aminoaldehydes produced by polyamine oxidation on the development *in vitro* of *Plasmodium falciparum* in normal and glucose-6-phosphate dehydrogenase deficient erythrocytes, *Biochem. J.*, 236, 97, 1986.

38. **Friedman, M. J.**, Control of malaria virulence by α1-acid glycoprotein (orosomucoid), an acute-phase (inflammatory) reactant, *Proc. Natl Acad. Sci. U.S.A.*, 80, 5421, 1983.

39. **Kharazmi, A. and Jepsen, S.**, Enhanced inhibition of *in vitro* multiplication of *Plasmodium falciparum* by stimulated human polymorphonuclear leucocytes, *Clin. Exp. Immunol.*, 57, 287, 1984.

40. **Farley, N. M., Shafer, W. M., and Spitznagel, J. K.**, Antimicrobial binding of a radiolabeled cationic neutrophil granule protein, *Infect. Immun.*, 55, 1536, 1987.

41. **Pontremoli, S., Melloni, E., Michetti, M., Sacco, O., Sparatore, B., Salamino, F., Damiani, G., and Horecker, B. L.**, Cytolytic effects of neutrophils: role for a membrane-bound neutral proteinase, *Proc. Natl. Acad. Sci. U.S.A.*, 83, 1685, 1986.

42. **Kharazmi, A., Jepsen, S., and Valerius, N. H.**, Polymorphonuclear leucocytes defective in oxidative metabolism inhibit *in vitro* growth of *Plasmodium falciparum, Scand. J. Immunol.*, 20, 93, 1984.

43. **Peto, T. E. A. and Thompson, J. L.**, A reappraisal of the effects of iron and desferrioxamine on the growth of *Plasmodium falciparum 'in vitro,'*: the unimportance of serum iron, *Br. J. Haematol.*, 63, 273, 1986.

44. **Harvey, P. W. J., Bell, R. G. and Nesheim, M. C.**, Iron deficiency protects inbred mice against infection with *Plasmodium chabaudi, Infect. Immun.*, 50, 932, 1985.

45. **Masawe, A. E. J., Muindi, J. M., and Swai, G. B. R.**, Infections in iron deficiency and other types of anaemia in the tropics, *Lancet,* ii, 1, 314, 1974.

46. **Murray, M. J., Murray, A. B., Murray, M. B., and Murray, C. J.**, The adverse effect of iron repletion on the course of certain infections, *Br. Med. J.*, ii, 1113, 1978.

47. **Fritsch, G., Sawatzki, G., Treumer, J., Jung, A., and Spira, D. T.**, *Plasmodium falciparum:* inhibition *in vitro* with lactoferrin, desferriferrithiocin, and desferricrocin, *Exp. Parasitol.*, 63, 1, 1987.

48. **Lelchuk, R., Taverne, J., Agomo, P. U., and Playfair, J. H. L.**, Development and suppression of a population of late-adhering macrophages in mouse malaria, *Parasite. Immunol.*, 1, 61, 1979.

49. **Brinkmann, V., Kaufmann, S. H. E., Simon, M. M., and Fischer, H.**, Role of macrophages during lethal *Plasmodium berghei* and self-limiting *Plasmodium yoelii* infection in mice, *Infect. Immun.*, 44, 743, 1984.

50. **Dockrell, H. M., Alavi, A., and Playfair, J. H. L.**, Changes in oxidative burst capacity during murine malaria and the effect of vaccination, *Clin. Exp. Immunol.*, 66, 37, 1986.

51. **Descamps-Latscha, B., Lunel-Fabiani, F., Karabinio, A., and Druilhe, P.**, Generation of reactive oxygen species in whole blood from patients with acute falciparum malaria, *Parasite Immunol.*, 9, 275, 1987.

52. **Taverne, J., Treagust, J. D., and Playfair, J. H. L.**, Macrophage cytotoxicity in lethal and non-lethal murine malaria and the effect of vaccination, *Clin. Exp. Immunol.*, 66, 44, 1986.

53. **Lucia, H. L. and Nussenzweig, R. S.**, *P. chabaudi* and *P. vinckei*: phagocytic activity of mouse reticuloendothelial system, *Exp. Parasitol.*, 25, 319, 1969.

54. **Shear, H. L., Nussenzweig, R. S., and Bianco, C.**, Immune phagocytosis in murine malaria, *J. Exp. Med.*, 149, 1288, 1979.

55. **Ward, K. N., Warrell, M. J., Rhodes, J., Looareesuwan, S., and White, N. J.**, Altered expression of human monocyte Fc receptors in *Plasmodium falciparum* malaria, *Infect. Immun.*, 44, 623, 1984.

56. **Lee, S.-H., Crocker, P., and Gordon, S.,** Macrophage plasma membrane and secretory properties in murine malaria, *J. Exp. Med.,* 163, 54, 1986.

57. **Taverne, J. Dockrell, H. M., and Playfair, J. H. L.,** Killing of the malarial parasite *Plasmodium yoelii in vitro* by cells of myeloid origin, *Parasite Immunol.,* 44, 77, 1982.

58. **Ockenhouse, C. F. and Shear, H. L.,** Malaria-induced lymphokines: stimulation of macrophages for enhanced phagocytosis, *Infect. Immun.,* 42, 73, 1983.

59. **Dale, D. C. and Wolff, S. M.,** Studies of the neutropenia of acute malaria, *Blood,* 41, 205, 1973.

60. **Brown, J. and Smalley, M. E.,** Inhibition of the *in vitro* growth of *Plasmodium falciparum* by human polymorphonuclear neutrophil leucocytes, *Clin. Exp. Immunol.,* 46, 106, 1981.

61. **Salmon, D., Vilde, J. L., Andrieu, B., Simonovic, R., and Lebras, J.,** Role of immune serum and complement in stimulation of the metabolic burst of human neutrophils by *Plasmodium falciparum, Infect. Immun.,* 51, 801, 1986.

62. **Kharazmi, A., Jepsen, S., and Andersen, B. J.,** Generation of reactive oxygen radicals of human phagocytic cells activated by *Plasmodium falciparum, Scand. J. Immunol.,* 25, 335, 1987.

63. **Trubowitz, S., and Masek, B.,** *Plasmodium falciparum:* phagocytosis by polymorphonuclear leucocytes, *Science,* 162, 273, 1968.

64. **Tosta, C. E. and Wedderburn, N.,** Immune phagocytosis of *Plasmodium yoelii*-infected erythrocytes by macrophages and eosinophils, *Clin. Exp. Immunol.,* 42, 114, 1980.

65. **Celada, A., Cruchaud, A., and Perrin, L. H.,** Phagocytosis of *Plasmodium falciparum*-parasitized erythrocytes by human polymorphonuclear leucocytes, *J. Parasitol.,* 69, 49, 1983.

66. **Weiss, S. J.,** Neutrophil-mediated methemoglobin formation in the erythrocyte. The role of superoxide and hydrogen peroxide, *J. Biol. Chem.,* 257, 2947, 1982.

67. **Butterworth, A. E., Remold, H. G., Houba, V., David, J. R., Franks, D., David, P. H., and Sturrock, R. F.,** Antibody dependent eosinophil-mediated damage to ^{51}Cr-labeled schistosomula of *Schistosoma mansoni* by IgG, and inhibition by antigen-antibody complexes, *J. Immunol.,* 118, 6, 1977.

68. **Wickramasinghe, S. N., Phillips, R. E., Looareesuwan, S., Warrell, D. A., and Hughes, M.,** Bone marrow in human cerebral malaria: parasite sequestration within sinusoids, *Br. J. Haematol.,* 66, 295, 1987.

69. **Zainal-Abidin, B. A. H., Robiah, Y., and Ismail, G.,** *Plasmodium berghei:* eosinophilic depression of infection in mice, *Exp. Parasitol.,* 57, 20, 1984.

70. **Theander, T. G., Pedersen, B. K., Bygbjerg, I. C., Jepsen, S., Larsen, P. B., and Kharazmi, A.,** Enhancement of human natural cytotoxicity by *Plasmodium falciparum* antigen activated lymphocytes, *Acta Trop.,* 44, 415, 1987.

71. **Ojo-Amaize, E. A., Salimonu, L. S., Williams, A. I. O., Akinwolere, O. A. O., Shabo, R., Alm, G. V., and Wigzell, H.,** Positive correlation between degree of parasitaemia, interferon titers, and natural killer cell activity in *Plasmodium falciparum*-infected children, *J. Immunol.,* 127, 2296, 1981.

72. **Eugui, E. M. and Allison, A. C.,** Differences in susceptibility of various mouse strains to haemoprotozoan infections: possible correlation with natural killer activity, *Parasite Immunol.,* 2, 277, 1980.

73. **Solomon, J. B.,** Natural cytotoxicity for *Plasmodium berghei in vitro* by spleen cells from susceptible and resistant rats, *Immunology,* 59, 277, 1986.

74. **Skamene, E., Stevenson, M. M., and Lemieux, S.,** Murine malaria: dissociation of natural killer (NK) cell activity and resistance to *P. chabaudi, Parasite Immunol.,* 5, 557, 1983.

75. **Wood, P. R. and Clark, I. A.,** Genetic control of *Propionbacterium acnes*-induced protection of mice against *Babesia microti, Infect. Immun.,* 35, 52, 1982.

76. **Metcalf, D.,** The role of the colony-stimulating factors in resistance to acute infection, *Immunol. Cell. Biol.,* 65, 35, 1987.

77. **Crawford, R. M., Finbloom, D. S., Ohara, J., Paul, W. E., and Meltzer, M. S.,** B cell stimulatory factor-1 (Interleukin 4) activates macrophages for increased tumoricidal activity and expression of Ia antigens, *J. Immunol.,* 139, 135, 1987.

78. **Magee, D. M., Wing, E. J., Ampel, N. M., Waheed, A., and Shadduck, R. K.,** Macrophage colony-stimulating factor enhances the expression of Fc receptors on murine peritoneal macrophages, *Immunology,* 62, 373, 1987.

79. **Guyre, P. M., Morganelli, P. M., and Miller, R.,** Recombinant immune interferon increases immuno-globulin G Fc receptors on cultured human mononuclear phagocytes, *J. Clin. Invest.,* 72, 393, 1983.

80. **Perussia, B., Dayton, E. T., Lazarus, R., Fanning, V., and Trinchieri, G.,** Immune interferon induces the receptor for monomeric IgG on human monocytic and myeloid cells, *J. Exp. Med.,* 158, 1092, 1983.

81. **Klempner, M. S., Dinarello, C. A., Henderson, W. R., and Gallin, J. I.,** Stimulation of neutrophil oxygen-dependent metabolism by human leukocytic pyrogen, *J. Clin. Invest.,* 64, 996, 1979.

82. **Esparza, I., Männel, D., Tuppel, A., Falk, W., and Krammer, P. H.,** Interferon γ and lymphotoxin or tumor necrosis factor act synergistically to induce macrophage killing of tumor cells and schistosomula of *Schistosoma mansoni, J. Exp. Med.,* 166, 589, 1987.

83. **Reed, S. G., Nathan, C. F., Pihl, D. L., Rodricks, P., Shanebeck, K., Conlon, P. J., and Grabstein, K. H.,** Recombinant granulocyte/macrophage colony stimulating-factor activates macrophages to inhibit *Trypanosoma cruzi* and release hydrogen peroxide. Comparison with Interferon γ, *J. Exp. Med.,* 166, 1734, 1987.

84. **Nathan, C. F., Murray, H. W., Wiebe, M. E., and Rubin, B. Y.,** Identification of gamma interferon as the lymphokine that activates human macrophage oxidative metabolism and antimicrobial activity, *J. Exp. Med.,* 158, 670, 1983.

85. **Murray, H. W., Spitalny, G. L., and Nathan, C. F.,** Activation of murine macrophages by interferon γ *in vitro* and *in vivo, J. Immunol.,* 134, 1619, 1985.

86. **Mokoena, T., and Gordon, S.,** Human macrophage activation, *J. Clin. Invest.,* 75, 624, 1985.

87. **Wing, E. J., Ampel, N. M., Waheed, A., and Shadduck, R. K.,** Macrophage colony-stimulating factor (M-CSF) enhances the capacity of murine macrophages to secrete oxygen reduction products, *J. Immunol.,* 135, 2052, 1985.

88. **Onazaki, K., Matsushima, K., Klienerman, E. S., Saito, T., and Oppenheim, J. J.,** Role of interleukin-1 in promoting human monocyte-mediated tumor cytotoxicity, *J. Immunol.,* 135, 314, 1985.

89. **Grabstein, K. H., Urdal, D. L., Tushinski, R. J., Mochizuki, D. Y., Price, V. L., Cantrell, M. A., Gillis, S., and Conlon, P. J.,** Induction of macrophage tumoricidal activity by granulocyte-macrophage colony-stimulating factor, *Science,* 232, 506, 1986.

90. **Pace, J. L., Russell, S. W., Torres, B. A., Johnson, H. M., and Gray, P. W.,** Recombinant mouse γ interferon induces the priming step in macrophage activation for tumor cell killing, *J. Immunol.,* 130, 2011, 1983.

91. **De Titto, E., Catteral, J. R., and Remington, J. S.,** Activity of recombinant tumor necrosis factor on *Toxoplasma gondii* and *Trypanosoma cruzi, J. Immunol.,* 137, 1342, 1986.

92. **Weiser, W. Y., van Niel, A., Clark, S. C., David, J. R., and Remold, J. G.,** Human biosynthetic granulocyte-macrophage colony stimulating factor (rGM-CSF) activates human macrophages to kill *Leishmania donovani, Fed. Am. Soc. Exp. Biol.,* 46, 6250, 1987.

93. **James, S. L.,** Activated macrophages as effector cells of protective immunity to schistosomiasis, *Immunol. Res.,* 5, 139, 1986.

94. **Ming, W. J., Bersani, L., and Mantovani, A.,** Tumor necrosis factor is chemotactic for monocytes and polymorpho-nuclear leukocytes, *J. Immunol.,* 138, 1469, 1987.

95. **Klebanoff, S. J., Vadas, M. A., Harlan, J. M., Sparks, L. H., Gamble, J. R., Agosti, J. M., and Waltersdorph, A. M.,** Stimulation of neutrophils by tumor necrosis factor, *J. Immunol.,* 136, 4220, 1986.

96. **Lopez, A. F., Williamson, D. J., Gamble, J. R., Begley, C. G., Harlan, J. M., Klebanoff, S. J., Waltersdorph, A., Wong, G., Clark, S. C., and Vadas, M. A.,** Recombinant human granulocyte-macrophage colony-stimulating factor stimulates *in vitro* mature human neutrophil and eosinophil function; surface receptor expression and survival, *J. Clin. Invest.,* 78, 1220, 1986.

97. **Tsujimoto, M., Yokota, S., Vilcek, J., and Weissmann, G.,** Tumor necrosis factor provokes superoxide anion generation from neutrophils, *Biochem. Biophys. Res. Commun.,* 137, 1094, 1986.

98. **Larrick, J. W., Graham, D., Toy, K., Lin, L. S., Senyk, G., and Fendly, B. M.,** Recombinant tumor necrosis factor causes activation of human granulocytes, *Blood,* 69, 640, 1987.

99. **Weisbart, R. H., Golde, D. W., Clark, S. C., Wong, G. G., and Gasson, J. C.,** Human granulocyte-macrophage colony-stimulating factor is a neutrophil activator, *Nature,* 314, 361, 1985.

100. **Shalaby, M. R., Aggarwal, B. B., Rinderknecht, E., Svedersky, L. P., Finkle, B. S., and Palladino, M. A., Jr.,** Activation of human polymorphonuclear neutrophil functions by interferon γ and tumor necrosis factors, *J. Immunol.,* 135, 2069, 1985.

101. **Djeu, J. Y., Blanchard, D. K., Halkias, D., and Friedman, H.,** Growth inhibition of *Candida albicans* by human polymorphonuclear neutrophils: activation by interferon-γ and tumor necrosis factor, *J. Immunol.,* 137, 2980, 1986.

102. **Vilalta, F. and Kierszenbaum, F.,** Effects of human colony-stimulating factor on the uptake and destruction of a pathogenic parasite *(Trypanosoma cruzi)* by human neutrophils, *J. Immunol.,* 137, 1703, 1986.

103. **Sauder, D. N., Monnessa, N. L., Katz, S. I., Dinarello, C. A., and Gallin, J. A.,** Chemotactic cytokines: the role of leukocytic pyrogen and epidermal cell thymocyte-activating factor in neutrophil chemotaxis, *J. Immunol.,* 132, 828, 1984.

104. **Lopez, A. F., To, L. B., Yang, Y.-C., Gamble, J. R., Shannon, M. F., Burns, G. F., Dyson, P. G., Juttner, C. A., Clark, S., and Vadas, M. A.,** Stimulation of proliferation, differentiation, and function of human cells by primate interleukin-3, *Proc. Natl. Acad. Sci. U.S.A.,* 84, 2761, 1987.

105. **Pincus, S. H., Whitcomb, E. A., and Cinarello, C. A.,** Interaction of IL-1 and TPA in modulating of eosinophil function, *J. Immunol.,* 137, 3509, 1986.

106. **Lopez, A. F., Begley, C. G., Williamson, D. J., Warren, D. J., Vadas, M. A., and Sanderson, C. J.,** Murine eosinophil differentiation factor. An eosinophil specific colony stimulating factor with activity for human cells, *J. Exp. Med.,* 163, 1085, 1986.

107. **Silberstein, D. S., and David, J. R.,** Tumor necrosis factor enhances eosinophil toxicity to *Schistosoma mansoni* larvae, *Proc. Natl. Acad. Sci. U.S.A.,* 83, 1055, 1986.

108. **Thorne, K. J. I., Richardson, B. A., Taverne, J., Williamson, D. J., Vadas, M. A., and Butterworth, A. E.,** A comparison of eosinophil-activating factor (EAF) with other monokines and lymphokines, *Eur. J. Immunol.,* 16, 1143, 1986.

109. **Silberstein, D. S. and David, J. R.,** The regulation of human eosinophil function by cytokines, *Immunol. Today,* 8, 380, 1987.

110. **Silberstein, D. S., Owen, W. F., Gasson, J. C., Di Persio, J. F., Golde, D. W., Bina, J. C., Soberman, R., Austen, K. F., and David, J. R.,** Enhancement of human eosinophil cytotoxicity and leukotriene synthesis by biosynthetic (recombinant) granulocyte-macrophage colony-stimulating factor, *J. Immunol.,* 137, 3290, 1986.

111. **Matsubara, T. and Ziff, M.,** Increased superoxide anion release from human endothelial cells in response to cytokines, *J. Immunol.,* 136, 329, 1986.

112. **Østensen, M. E., Thiele, D. L., and Lipsky, P. E.,** Tumor necrosis factor-α enhances cytolytic activity of human natural killer cells, *J. Immunol.,* 138, 4185, 1987.

113. **Schofield, L., Villaquiran, J., Ferreira, A., Schellekens, H., Nussenzweig, R. S., and Nussenzweig, V.,** γ Interferon, CD8 + T cells and antibodies required for immunity to malaria sporozoites, *Nature,* 330, 664, 1987.

114. **Ojo-Amaize, E. A., Vilček, J., Cochrane, A. H., and Nussenzweig, R. S.,** *Plasmodium berghei* sporozoites are mitogenic for murine T cells, induce interferon, and activate killer cells, *J. Immunol.,* 133, 1005, 1984.

115. **Wyler, D. J., Herrod, H. G., and Weinbaum, F. I.,** Response of sensitised and unsensitised human lymphocyte sub-populations to *Plasmodium falciparum* antigens, *Infect. Immun.,* 24, 106, 1979.

116. **Ballet, J. J., Druilhe, P., Querleux, M. A., Schmitt, C., and Agrapart, M.,** Parasite-derived mitogenic activity for human T cells in *Plasmodium falciparum* continuous cultures, *Infect. Immun.,* 33, 758, 1981.

117. **Gabrielsen, A. A. and Jensen, J. B.,** Mitogenic activity of extracts from continuous cultures of *Plasmodium falciparum, Am. J. Trop. Med. Hyg.,* 31, 441, 1982.

118. **Kataaha, P. K., Facer, C. A., Mortazavi-Milani, S. M., Stievle, H., and Holborow, E. J.,** Stimulation of autoantibody production in normal blood lymphocytes by malaria culture supernatants, *Parasite Immunol.,* 6, 481, 1984.

119. **Theander, T. G., Bygbjerg, I. C., Jepsen, S., Svenson, M., Kharazmi, A., Larsen, P. B., and Bendtzen, K.,** Proliferation induced by *Plasmodium falciparum* antigen and interleukin-2 production by lymphocytes isolated from malaria-immune individuals, *Infect. Immun.,* 53, 221, 1986.

120. **Ho, M., Webster, H. K., Looareesuwan, S., Supanavanond, W., Phillips, R. E., Chanthavanich, P., and Warrell, D. A.,** Antigen-specific immunosuppression in human malaria due to *Plasmodium falciparum, J. Infect. Dis.,* 153, 763, 1986.

121. **Lelchuk, R., Dockrell, H. M., and Playfair, J. H. L.,** T independent macrophage changes in murine malaria, *Clin. Exp. Immunol.,* 51, 487, 1983.

122. **Dotsika, E. N. and Sanderson, C. J.,** Interleukin-3 production as a sensitive measure of T-lymphocyte activation in the mouse, *Immunology,* 62, 665, 1987.

123. **Holter, W., Goldman, C. K., Casabo, L., Nelson, D. L., Greene, W. C., and Waldmann, T. A.,** Expression of functional IL-2 receptors by lipopolysaccharide and interferon γ stimulated human monocytes, *J. Immunol.,* 138, 2917, 1987.

124. **Strickland, G. T., Ahmed, A., Sayles, P. C., and Hunter, K. W.,** Murine malaria: cellular interactions in the immune response, *Am. J. Trop. Med. Hyg.,* 32, 1229, 1983.

125. **Fahey, J. R. and Spitalny, G. L.,** Rapid parasite multiplication rate, rather than immunosuppression, causes the death of mice infected with virulent *Plasmodium yoelii, Infect. Immun.,* 55, 490, 1987.

126. **Roberts, D. W. and Weidanz, W. P.,** Splenomegaly, enhanced phagocytosis and anemia are thymus-dependent responses to malaria, *Infect. Immun.,* 20, 728, 1978.

127. **Jayawardena, A. N., Targett, G. A. T., Carter, R. L., Leuchars, E., and Davies, A. J. S.,** The immunological response of CBA mice to *P. yoelii.* I. General characteristics, the effects of T cell deprivation and reconstitution with thymus grafts, *Immunology,* 32, 849, 1977.

128. **Wyler, D. J. and Gallin, J. I.,** Spleen-derived mononuclear cell chemotactic factor in murine malaria infection: a possible mechanism for splenic macrophage accumulation, *J. Immunol.,* 118, 478, 1977.

129. **Dockrell, H. M., de Souza, J. B., and Playfair, J. H. L.,** The role of the liver in immunity to blood-stage murine malaria, *Immunology,* 41, 421, 1980.

130. **Taverne, J., Rahman, D., Dockrell, H. M., Alavi, A., Leveton, C., and Playfair, J. H. L.,** Activation of liver macrophages in murine malaria is enhanced by vaccination, *Clin. Exp. Immunol.,* 70, 508, 1987.

131. **Mogil, R. J., Patton, C. L., and Green, D. R.,** Cellular subsets involved in cell-mediated immunity to murine *Plasmodium yoelii* 17X malaria, *J. Immunol.,* 138, 1933, 1987.

132. **Troye-Blomberg, M., Andersson, G., Stoczkowska, M., Shabo, R., Romero, P., Patarroyo, E., Wigzell, H., and Perlmann, P.,** Production of IL-2 and IFN-γ by T cells from malaria patients in response to *Plasmodium falciparum* or erythrocyte antigens *in vitro, J. Immunol.,* 135, 3498, 1985.

133. **White, N. J. and Dance, D. A. B.,** Clinical and laboratory studies of malaria and meioidosis, *Trans. R. Soc. Trop. Med. Hyg.,* 82, 15, 1988.

134. **Lelchuk, R., Rose, G., and Playfair, J. H. L.,** Changes in the capacity of macrophages and T cells to produce interleukins during murine malaria infection, *Cell. Immunol.,* 84, 253, 1984.

135. **Lelchuk, R. and Playfair, J. H. L.,** Serum IL-2 inhibitor in mice. I. Increase during infection, *Immunology,* 56, 113, 1985.

136. **Male, D., Lelchuk, R., Curry, S., Pryce, G., and Playfair, J. H. L.,** Serum IL-2 inhibitor in mice. II. Molecular characteristics, *Immunology,* 56, 119, 1985.

137. **Havell, E. A., Spitalny, G. L., and Patel, P. J.,** Enhanced production of murine interferon-γ by T cells generated in response to bacterial infection, *J. Exp. Med.,* 156, 112, 1982.

138. **Rhodes-Feuillette, A., Bellosguardo, M., Druilhe, P., Ballet, J. J., Chousterman, S., Canivet, M., and Peries, J.,** The interferon compartment of the immune response in human malaria. II. Presence of serum-interferon gamma following the acute attack, *J. Interferon Res.,* 5, 169, 1985.

139. **Huang K.-Y., Schultz, W. W., and Gordon, F. B.,** Interferon induced by *Plasmodium berghei, Science,* 162, 123, 1968.

140. **Rönnblom, L., Ojo-Amaize, E. A., Franzen, L., Wigzell, H., and Alm, G. V.,** *Plasmodium falciparum* parasites induce interferon production on human peripheral blood "null" cells *in vitro, Parasite Immunol.,* 5, 165, 1983.

141. **Jahiel, R. I., Nussenzweig, R. S., Vanderberg, J., and Vilček, J.,** Anti-malarial effect of interferon inducers at different stages of development of *Plasmodium berghei* in the mouse, *Nature,* 220, 710, 1968.

142. **Jahiel, R. I., Vilcek, J., and Nussenzweig, R. S.,** Exogenous interferon protects mice against *Plasmodium berghei* malaria, *Nature,* 227, 1350, 1970.

143. **Sauvager, F. and Fauconnier, B.,** The protective effect of endogenous interferon in mouse malaria, as demonstrated by the use of anti-interferon globulins, *Biomedicine,* 29, 184, 1978.

144. **Playfair, J. H. L.,** Immunity to malaria, *Br. Med. Bull.,* 38, 153, 1982.

145. **Ferreira, A., Schofield, L., Enea, V., Schellekens, H., Van der Meide, P., Collins, W. E., Nussenzweig, R. S., and Nussenzweig, V.,** Inhibition of development of exoerythrocytic forms of malaria parasites by γ-interferon, *Science,* 232, 881, 1986.

146. **Maheshwari, R., Czarniecki, C. W., Dutta, G. P., Puri, S. K., Dharwan, B. N., and Friedman, R.,** Recombinant human gamma interferon inhibits simian malaria, *Infect. Immun.,* 53, 628, 1986.

147. **Ferreira, A., Enea, V., Morimoto, T., and Nussenzweig, V.,** Infectivity of *Plasmodium berghei* sporozoites measured with a DNA probe, *Mol. Biochem. Parasitol.,* 19, 103, 1986.

148. **Schofield, L., Ferreira, A., Altszuler, R., Nussenzweig, V., and Nussenzweig, R. S.,** Interferon γ inhibits the intrahepatocytic development of malaria parasites *in vitro, J. Immunol.,* 139, 2020, 1987.

149. **Vergara, U., Ferreira, A., Schellekens, H., and Nussenzweig, V.,** Mechanism of escape of exoerythrocytic forms (EEF) of malaria parasites from the inhibitory effects of interferon γ, *J. Immunol.,* 138, 4447, 1987.

150. **Mellouk, S., Maheshwari, R. K., Rhodes-Feuillette, A., Beaudoin, R. L., Berbiguier, N., Hughes, M., Miltgen, F., Landau, I., Pied, S., Chigot, J. P., Friedman, R. M. and Mazier, D.,** Inhibitory activity of interferons and interleukin 1 on the development of *Plasmodium falciparum* in human hepatocyte cultures, *J. Immunol.,* 139, 4192, 1987.

151. **Jensen, J. B., Boland, M. T., Allan, J. S., Carlin, J. M., Van de Waa, J. A., Divo, A. A., and Akood, M. A. S.,** Association between human serum-induced crisis forms in cultured *Plasmodium falciparum* and clinical immunity to malaria in Sudan, *Infect. Immun.,* 41, 1302, 1983.

152. **Carlin, J. M., Jensen, J. B., and Geary, T. G.,** Comparison of inducers of crisis forms in *Plasmodium falciparum in vitro, Am. J. Trop. Med. Hyg.,* 34, 668, 1985.

153. **Ockenhouse, C. F. and Shear, H. L.,** Oxidative killing of the intraerythrocytic malaria parasite *Plasmodium yoelii* by activated macrophages, *J. Immunol.,* 132, 424, 1984.

154. **Le, J. and Vilček, J.,** Biology of disease. Tumor necrosis factor and interleukin 1: cytokines with multiple overlapping biological activity, *Lab. Invest.,* 56, 234, 1987.

155. **Collart, M. A., Bellin, D., Vassali, J.-D., De Kassado, S., and Varsalli, P.,** γ-Interferon enhances macrophage transcription of the tumor necrosis factor/cachectin, interleukin-1 and urokinase genes, which are controlled by short-lived repressors, *J. Exp. Med.,* 164, 2113, 1986.

156. **Clark, I. A.,** Cell-mediated immunity in protection and pathology of malaria, *Parasitol. Today,* 3, 300, 1987.

157. **Zucali, J. R., Elfenbeim, G. J., Barth, K. C., and Dinarello, C. A.,** Effects of human interleukin 1 and human tumor necrosis factor on human T lymphocyte colony formation, *J. Clin. Invest.,* 80, 772, 1987.

158. **Scheurich, P., Thoma, B., Ücer, U., and Pfizenmaier, K.,** Immunoregulatory activity of recombinant human tumor necrosis factor (TNF)-α: induction of TNF receptors on human T cells and TNF-α-mediated enhancement of T cell responses, *J. Immunol.,* 138, 1786, 1987.

159. **Yokota, S., Geppert, T. D., and Lipsky, P.,** Enhancement of antigen- and mitogen-induced human T lymphocyte proliferation by tumor necrosis factor, *J. Immunol.,* 140, 531, 1988.

160. **Le, J., Weinstein, D., Gubler, U., and Vilček, J.,** Induction of membrane-associated interleukin 1 by tumor necrosis factor in human fibroblasts, *J. Immunol.,* 138, 2137, 1987.

161. **Ruggiero, V. and Baglioni, C.,** Synergistic anti-proliferative activity of interleukin-1 and tumor necrosis factor, *J. Immunol.,* 138, 661, 1987.

162. **Elias, J. A., Gustilo, K., Baeder, W., and Freundlich, B.,** Synergistic stimulation of fibroblast prostaglandin production by recombinant interleukin-1 and tumor necrosis factor, *J. Immunol.,* 138, 3812, 1987.

163. **Holtmann, H. and Wallach, D.,** Down regulation of the receptors of tumor necrosis factor by interleukin 1 and 4 β-phorbol-12-myristate-13-acetate, *J. Immunol.,* 139, 1161, 1987.

164. **Movat, H. Z.,** Tumor necrosis factor and interleukin-1; role in acute inflammation and microvascular injury, *J. Lab. Clin. Med.,* 110, 668, 1987.

165. **Wood, P. R. and Clark, I. A.,** Macrophages from *Babesia* and malaria-infected mice are primed for monokine release, *Parasite Immunol.,* 6, 309, 1984.

166. **Spira, D. T., Golenser, J., and Gery, I.,** The reactivity of spleen cells from malarious rats to non-specific mitogens, *Clin. Exp. Immunol.,* 24, 139, 1976.

167. **Wyler, D. J., Oppenheim, J. J., and Koontz, L. C.,** Influence of malaria infection on the elaboration of soluble mediators by adherent mononuclear cells, *Infect. Immun.,* 24, 151, 1979.

168. **Scuderi, P., Sterling, K. E., Lam, K. S., Finley, P. R., Ryan, K. J., Ray, C. G., Petersen, E., Slymen, D., and Salmon, S. E.,** Raised serum levels of tumour necrosis factor in parasitic infections, *Lancet,* ii, 1364, 1986.

169. **Grau, G. B., Fajardo, L. F., Piguet, P.-F., Allen B., Lambert, P.-H., and Vassalli, P.,** Tumor necrosis factor (cachectin) as an essential mediator in murine cerebral malaria, *Science,* 237, 1210, 1987.

170. **Clark, I. A., Virelizier, J.-L., Carswell, E. A., and Wood, P. R.,** Possible importance of macrophage-derived mediators in acute malaria, *Infect. Immun.,* 32, 1058, 1981.

171a. **Bate, C. A. W., Taverne, J., and Playfair, J. H. L.,** Malarial parasites induce TNF production by macrophages, *Immunology,* 64, 227, 1988.

171b. **Bate, C. A. W., Taverne, J., and Playfair, J. H. L.,** Soluble malarial antigens are toxic and induce production of tumour necrosis factor *in vivo, Immunology,* 66, 600, 1989.

171c. **Taverne, J., Bate, C. A. W., Sarkar, D. A., Meager, A., Rook, G. A. W., and Playfair, J. H. L.,** Human and murine macrophages produce TNF in response to soluble antigens of *P. falciparum,* submitted.

172. **Jakobsen, P. H., Baek, L., and Jepsen, S.,** Demonstration of soluble *Plasmodium falciparum* antigens reactive with Limulus amoebocyte lysate and polymyxin B, *Parasite Immunol.,* 10, 593, 1988.

173. **Lehmann, V., Freudenberg, M. A., and Galanos, C.,** Lethal toxicity of lipopolysaccharide and tumor necrosis factor in normal and D-galactosamine-treated mice, *J. Exp. Med.,* 165, 657, 1987.

174. **Nussenzweig, V.,** in *Tumor Necrosis Factor and Related Cytotoxins,* Ciba Foundation Symp. 131, John Wiley & Sons, New York, 1987, 198.

175. **Clark, I. A., Hunt, N. H., Butcher, G. A., and Cowden, W. B.,** Inhibition of murine malaria (*Plasmodium chabaudi*) *in vivo* by recombinant interferon-γ or tumor necrosis factor and its enhancement by butylated hydroxyanisole, *J. Immunol.,* 139, 3493, 1987.

176. **De Souza, J. B. and Playfair, J. H. L.,** Antilymphocytic autoantibody in lethal mouse malaria and its suppression by non-lethal malaria, *Parasite Immunol.,* 5, 257, 1983.

177. **Playfair, J. H. L. and de Souza, J. B.,** Lymphocyte traffic and lymphocyte destruction in murine malaria, *Immunology,* 46, 125, 1982.

178. **Lelchuk, R. and Playfair, J. H. L.,** Two distinct types of non-specific immunosuppression in murine malaria, *Clin. Exp. Immunol.,* 42, 448, 1980.

179. **Greenwood, B. M., Playfair, J. H. L., and Torrigiani, G.,** Immunosuppression in murine malaria. I. General characteristics, *Clin. Exp. Immunol.,* 8, 467, 1978.

180. **Weidanz, W. P.,** Malaria, and alterations in immune reactivity, *Br. Med. Bull.,* 38, 167, 1982.

181. **Ballet, J. J., Druilhe, P., Vasconcelos, I., Schmitt, C., Agrapart, M., and Frommel, D.,** Human lymphocyte responses to *Plasmodium falciparum* antigens. A functional assay of protective immunity?, *Trans R. Soc. Trop. Med. Hyg.,* 79, 497, 1985.

182. **Theander, T. G., Bygbjerg, I. C., Jacobsen, L., Jepsen, S., Larsen, P. B., and Kharazmi, A.,** Low parasite specific T cell response in clinically immune individuals with low grade *Plasmodium falciparum* parasitaemia, *Trans. R. Soc. Trop. Med. Hyg.,* 80, 1000, 1986.

183. **Jayawardena, A. N., Murphy, D. B., Janeway, C. A., and Gershon, R. K.,** T cell-mediated immunity to malaria. I. The Ly phenotype of T cells mediating resistance to *Plasmodium yoelii, J. Immunol.,* 34, 377, 1982.

184. **De Souza, J. B. and Playfair, J. H. L.,** Immunization of mice against blood-stage *Plasmodium yoelii* malaria with isoelectrically focussed antigens and correlation of immunity with T cell priming *in vivo, Infect. Immun.,* 56, 88, 1988.

185. **Sinigaglia, F. and Pink, J. R. L.,** Human T lymphocyte clones specific for malaria (*Plasmodium falciparum*) antigens, *EMBO J.,* 4, 3819, 1985.

186. **Abou-Zeid, C., Filley, E., Steele, J., and Rook, G. A. W.,** A simple new method for using antigens separated by polyacrylamide gel electrophoresis to stimulate lymphocytes *in vitro* after converting bands cut from Western blots into antigen-bearing particles, *J. Immunol. Methods.,* 98, 5, 1987.

Chapter 4

THE ROLE OF MACROPHAGES IN RESISTANCE TO MALARIA

Hannah Lustig Shear

TABLE OF CONTENTS

I. INTRODUCTION

Macrophages play a central role in the immune response. They, along with other antigen-presenting cells, process and present antigen to T cells initiating the immune response.[1] Macrophages are not only the major phagocytic cell but are also involved in the synthesis and secretion of at least 100 products, including reactive oxygen metabolites, polypeptide hormones, enzymes and enzyme inhibitors, complement components, and many other important proteins and chemicals.[2] When appropriately stimulated, macrophages are cytotoxic to tumor cells.[3] Of obvious relevance to this chapter is the role of macrophages in the destruction of microorganisms, particularly parasites.

The ability of monocytes and macrophages to kill protozoan organisms has been most extensively studied with Leishmania parasites and *Toxoplasma gondii*,[4] although it is clear that other organisms such as *Trypanosoma cruzi*[5] are also susceptible to killing by appropriately activated macrophages.

A large body of information has led to several major findings on the killing of intracellular protozoan parasites. As reviewed by Murray,[4] fresh peripheral blood monocytes can kill about 90% of *Leishmania donovani* promastigotes and 30 to 40% of the amastigotes of this parasite, and about 40% of the trophozoites of *T. gondii, in vitro*. Cytotoxicity against these parasites is correlated with the oxidative burst capacity of the monocytes. In contrast, monocyte-derived macrophages do not kill these parasites but allow them to multiply. However, the ability of macrophages to inhibit the growth of these parasites or kill them can be restored by treating them with crude T cell-derived lymphokines. The active component of the lymphokines is γ-interferon. The capacity of monocytes and lymphokine or γ-interferon-activated macrophages to kill *Leishmania*, for example, is associated with both oxygen-dependent and -independent mechanisms.[4] The nature of the oxygen-independent mechanism is as yet unknown.

As will be discussed in this chapter, monocytes and activated macrophages may exert similar cytotoxic effects on the intraerythrocytic malaria parasite.

A role for macrophages in the host response to malaria was proposed over 100 years ago by a list of notable medical scientists including Metchnikoff, Marchiafava and Celli, Golgi, and Laverin.[6] Experimental data on how macrophages participate in the host response were first obtained by Taliaferro and Cannon[7,8] in studies using canaries and monkeys as experimental hosts. Their observations, which were mainly histologic, indicated that in experimental infections there was an initial rise in parasitemia during which parasites were slowly phagocytized, primarily in the spleen and to a lesser extent in the liver and bone marrow. As the animals approached peak parasitemia, a "crisis" occurred in which the parasitemia dropped sharply and the remaining parasites appeared pyknotic and were called "crisis forms". The parasitized erthrocytes appeared to be held in the cords of Billroth in the spleen. Within the next day or two, the parasites were ingested more aggressively and the infection was partially or completely cleared. Upon reinfection, parasitized erythrocytes were phagocytosed within hours rather than days, indicating that acquired immunity to infection could augment the capacity of macrophages to ingest parasites. While the target of the splenic macrophage appeared to be the infected erythrocyte, free parasites were probably also ingested but destroyed more rapidly and therefore difficult to identify microscopically. In addition, in some primate malarias, crisis also resulted in changes in the length and synchrony of the malaria cycle as well as in the number of merozoites released per schizont.[9]

The findings of Taliaferro and collaborators have recently been extended by Weiss, et al.[10] in an ultrastructural study of the spleen during nonlethal and lethal *Plasmodium yoelii* 17x infections of mice. These investigators have shown that there is a progressive change in the spleen during malaria involving the activation of reticular cells and the closing of a blood-spleen barrier in the precrisis phase. This, the investigators suggest, serves to protect the developing erythrocytes and lymphoid cells. At crisis the barrier opens, parasitized erythrocytes are held by the reticular cells and actively phagocytized by the macrophages. During infection with the lethal variant of *P. yoelii* 17x, the barrier is not formed and erythrocytes are continually parasitized and phagocytized until the host is overwhelmed. These findings raise fascinating questions as to the mechanisms of activation of the reticular cells and the nature of their secretory products.

The protective effects of macrophages, however, have to be reconciled with their well-established role in contributing to the anemia of malaria, which is due to the loss of both parasitized as well as nonparasitized erythrocytes. In a series of studies in rats infected with *P. berghei*, Zuckerman et al.[11,12] showed that the uptake of red cells per splenic macrophage was greatly increased over normal spleen macrophages (up to 200-fold), particularly around the time of peak parasitemia. However, greater numbers of uninfected than infected erythrocytes were taken up. She suggested that antibodies or immune complexes sensitized uninfected erythrocytes and mediated their ingestion by macrophages. Later work proved her hypothesis to be correct, since IgG and/or complement components were found on erythrocytes of both experimentally infected rodents[13,14] and humans living in malaria endemic regions.[15,16] It has also been found that this sensitization correlates with anemia in infected children.[17] However, Abdalla and Weatherall[18] did not find a correlation between anemia and the presence of antibodies on erythrocytes of children with malaria. This may reflect differences in the populations studied or in the methodologies utilized.

At least three basic questions remain in relation to the role of macrophages in malaria: (1) How do macrophages damage intraerythrocytic malaria parasites?, (2) Why are macrophages not always successful at eradicating malarial infection?, and (3) How can the parasiticidal role of macrophages be augmented in prophylaxis and/or treatment of malaria, without causing damage to the host?

It is the purpose of this chapter to review studies from this laboratory as well as the literature on the role of macrophages in the immune response to malaria which may help answer these questions and lead to further experimentation in this area.

II. CHANGES IN MACROPHAGES DURING MALARIA INFECTION

It is well known that the spleen is greatly enlarged during malaria. Indeed, the spleen rate is a measure of malaria endemicity.[19] This enlargement is due, at least partially, to the enormous increase in the number of macrophages in the spleen.[6,20] Further, the clearance of carbon particles is greatly increased during experimental malaria.[21] However, these findings do not distinguish between the possibilities that the macrophages are activated or that there are simply more of them.

Therefore, several years ago, we initiated studies on the functional activities of macrophages during malaria.[22] SW mice were infected with 10,000 *Plasmodium berghei*-infected erthrocytes intravenously. At various times after infection animals were sacrificed, peritoneal and splenic macrophages were isolated, and macrophages were assayed for ability to spread on glass for their Fc and C3 receptor activity.

We found that the size of splenic macrophages increased dramatically during the course of malarial infection. Splenic macrophages from uninfected mice had an average diameter of 12.5 μm. During malaria infection, the size of splenic macrophages increased to 22.9 μm (day 7), 32.9 μm (day 14), and 39.1 μm (day 21). In addition, the percentage of macrophages that spread rapidly on glass increased from 21.0% in uninfected mice to 52.2% on day 21 of *P. berghei* infection.

The effect of malaria on Fc and C3 receptor activity is shown in Figure 1. We found

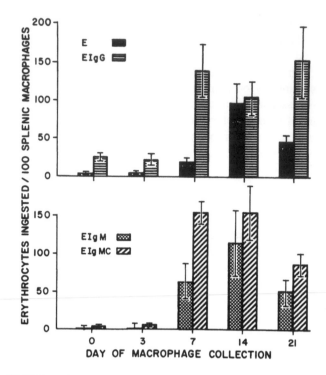

FIGURE 1.   Phagocytosis of E, EIgG, EIgM, and EIgMC (sheep erythrocytes unsensitized or sensitized with IgG, IgM, or IgM plus complement,
respectively) by spleen macrophages obtained from *P. berghei*-infected
mice. Macrophages were obtained from normal (day 0) or infected mice
on days 3, 7, 14, and 21 of infection. They were incubated with the various
erythrocyte preparations for 45 min. Ingestion was determined by counting
200 macrophages on duplicate cover slips. Data are expressed as the mean
number of erythrocytes ingested/100 macrophages ± SE. Groups of three
to nine animals were used per point. ■, E; □, EIgG; ▨ , EIgM; and
▨ EIgMC. (From Shear, H. L., Nussenzweig, R. S., and Bianco, C., *J.
Exp. Med.*, 163, 54, 1986. With permission.)

ingestion of IgG-sensitized cells as well as binding of C3b-coated cells by splenic macrophages from infected mice greatly increased, compared to splenic macrophages from normal
mice, from day 7 of infection on. There was even enhanced ingestion of control particles
(unsensitized E) which we found to be due to the synthesis of anti-Forssmann antibodies
during this infection. In addition splenic macrophages from infected mice ingested *P. berghei*-
infected erythrocytes but did not ingest normal mouse erythrocytes or reticulocytes. This
ingestion appeared to be mediated by erythrocyte-associated antibodies since pretreatment
of the erythrocytes with protein A reduced their ingestion. Interestingly, peritoneal macrophages of malaria-infected mice were not activated, suggesting that the effect was localized
to the spleen and perhaps the liver.

A more recent study by Lee et al.[23] looked at the antigens, receptors, and secretory
properties of spleen, liver, and bone marrow macrophages during *P. yoelii* 17x infection in
mice, which is not lethal. Using a monoclonal antibody against the macrophage-specific
membrane antigen F4/80, a fourfold increase in this antigen was found per unit volume of
liver tissue and a sevenfold total increase was seen. There was also a 13-fold increase in
the amount of F4/80 in spleen macrophages, as measured by immunoprecipitation. The data
indicated that the increase in macrophages was due to newly recruited macrophages from
monocytes in the blood since many of the liver macrophages expressed the F7/4 antigen,
not normally present on Kupffer cells. Functional changes in liver macrophages were also

found. These macrophages had increased size and ability to spread, increased phagocytic activity, and respiratory burst activity.

III. CHANGES IN THE ERYTHROCYTE MEMBRANE DURING MALARIA INFECTION

What is the evidence that malaria-parasitized erythrocytes have surface alterations which allow them to be recognized by antibodies or immune effector cells such as activated macrophages? Many studies now show that the erythrocyte surface is altered during malaria and several malarial antigens are expressed on the red blood cell. These will be mentioned briefly since they may be the target of antibodies and macrophage activity.

Changes in the erythrocyte membranes of malaria-infected erythrocytes were first observed at the electron microscopic level by Rudzinska and Trager[24] on *P. coatneyi*-infected Rhesus erythrocytes and by Luce and Miller[25] and Miller[26] on *P. falciparum*-infected red blood cells of Aotus and man, respectively. The most prominent change was the appearance of the so-called "knob", an electron-dense excrescence lying under the red cell membrane. The knobs are thought to play a role in the sequestration of mature forms of the asexual stages by causing the erythrocytes containing these parasites to adhere to the endothelium of the internal organs, including the brain.

The fact that the knobs were antigenic was demonstrated by Langreth and Reese.[27] The knobs consist of electron-dense cups which contain an unusual histidine-rich protein.[28,29] The cups and the histidine-rich protein appear to be anchored to the erythrocyte skeleton.[28]

Other antigenic molecules, in addition to the histidine-rich protein, appear to be involved in the cytoadherance of "knobby" erythrocytes to the endothelial membranes of the capillaries of the internal organs. This has been studied using human endothelial cells[30] and an amelanotic melanoma cell[31] line which binds the late stages of *P. falciparum*-infected K+ (knob+) cells, but not K− cells, via their knobs. Interestingly, immune serum can reverse the binding of the parasitized erythrocytes to the melanoma cells,[32] suggesting that the cytoadherance molecule(s) are antigenic. Although these molecules appear to be strain specific,[32,33] they may also contain conserved domains.[34]

Another antigen which is of interest is the variant antigen on the surface of malaria-infected erythrocytes. This was first observed in *P. knowelsi*[35] and then on *P. falciparum*.[36] Although this antigen may not be a suitable target for vaccination, it is of interest because of the control of its expression by the spleen.[36-40] There is no information as yet on how the spleen controls the expression of this antigen or whether macrophages play any role in this.

An antigen, expressed on the surface of malaria-infected erythrocytes, which has received considerable attention as a vaccine candidate is Pf155[41] or RESA.[42] This antigen is thought to be secreted by the rhoptries and deposited on the membrane of infected erythrocytes during invasion. Immune sera from people living in malaria-endemic areas which have antiPf155 activity inhibit the invasion of erythrocytes by *P. falciparum*.[43] The unusual feature of this antigen is that the erythrocytes must be fixed with gluteraldehyde in order to detect it.[41] Therefore, the antigen may not be exposed on the red cell surface. Nevertheless, this antigen has been cloned and peptides derived from these clones used as immunogens in vaccine trials.[44] Some peptides, containing repetitive epitopes, were found to be partially protective. Further, the protection was associated with the presence of antibodies to the peptide used. Since the partially protected animals did experience a course of parasitemia, we do not know what the precise mechanism of protection against this antigen is.

One way to approach this question is the use of an animal model which can be more easily manipulated. For this reason we recently immunized mice with the erythrocyte membranes of *P. chabaudi*-infected red blood cells.[45] Sera and monoclonal antibodies obtained from the immunized mice reacted with *P. chabaudi*-infected erythrocytes in the same way as antisera against Pf155 or RESA reacted with *P. falciparum*; that is, we obtained a perimeter flourescence around gluteraldehyde-fixed, infected erythrocytes. However, in the *P. chabaudi* system all maturational stages of the intraerythrocytic parasites were reactive whereas antibodies against Pf155 or RESA primarily stain the ring stages of *P. falciparum*. Perlmann's group has also identified the same antigen in *P. chabaudi*, however, their monoclonal detects only early erythrocytic stages.[46]

Our monoclonal antibodies precipitated a protein of Mr 96 kDa.[47] We used the monoclonal antibodies to purify the 96 kDa protein and immunized mice with the column-purified protein. Mice immunized with this protein, together with saponin as an adjuvant, were partially protected from challenge with *P. chabaudi*. Current studies are aimed at determining the mechanism of protection in this system. The gene for this protein has been cloned[48] and we may be able to use protein synthesized *in vitro* for these studies. This will allow us to determine whether macrophages play a role in this protection. Interestingly, the monoclonals against Pc96 cross-react with ring stages and more mature blood stages of *P. falciparum*, *P. vivax,* and *P. cynomolgi* (Wanidworanun, C. et al., in press). Therefore, Pc96 may be a functionally important protein. In addition, the information we obtain on immunization may be applicable to human vaccination.

Finally a protein has been identified on the erythrocyte membrane of *P. falciparum*-infected erythrocytes which may function as a transferrin receptor.[49,50] Whether antibodies or cellular mechanisms against this protein play a role in immune resistance to malaria is yet to be determined.

IV. FUNCTIONAL PROPERTIES OF MACROPHAGES DURING MALARIA

A. PHAGOCYTOSIS

The initial studies on the role of macrophages in malaria emphasized the phagocytic capacity of these cells. What is the evidence that phagocytosis of parasitized erythrocytes and/or free parasites occurs during malaria and that it is an important aspect of the immune response to malaria?

Based on the findings of Taliaferro and collaborators, many investigators attempted to determine whether phagocytosis of parasitized erythrocytes occurred in vitro with mixed results. When considering these data, it is important to bear in mind that the ability of macrophages to ingest a particle *in vitro* cannot be identical to its ability to ingest a particle *in vivo* since the architecture of the organ from which the macrophage was derived and the surrounding microenvironment cannot be duplicated *in vitro*.

Nevertheless, several investigators have observed ingestion of malaria-parasitized erythrocytes *in vitro*. One of the first of these studies was that of Brown.[51] Peritoneal macrophages obtained from BCG-infected mice were found to ingest *P. knowelsi*-infected red cells more effectively than did normal mouse macrophages. In addition, opsonizing serum was necessary for this ingestion. In a homologous rodent system, Tosta and Wedderburn[52] observed ingestion of late developmental stages of *P. yoelii*-infected erythrocytes by unstimulated mouse peritoneal macrophages. Once again, immune serum greatly enhanced ingestion. In this study, macrophages could discriminate between parasitized and nonparasitized erythrocytes. Finally, ingestion of the human malaria parasite has been observed *in vitro*. Using monocytes from normal volunteers, Celada and co-workers[53] observed ingestion of mature forms of *P. falciparum*-infected erythrocytes but no substantial ingestion of normal erythrocytes. Furthermore, serum from malaria-immune individuals (and purified IgG) greatly enhanced the ingestion of infected erythrocytes.

On the other hand, Khusmuth, et al.[54] did not observe ingestion of *P. falciparum*-infected erythrocytes by normal or malarial subjects; however, no immune serum was included in their assay. Chow and Kreier[55] also did not observe substantial ingestion of *P. berghei*-parasitized erythrocytes *in vitro* in the presence of immune serum, although ingestion of merozoites and trophozoites was observed.[55-57] Although it is not clear why no ingestion of infected erythrocytes was seen, these investigators used unstimulated macrophages which would not have given optimal results. In addition, since *P. berghei* is not a synchronous parasite, many of the target erythrocytes would have contained young parasites and probably would not have expressed surface alterations on the red cell membrane. As mentioned above, we were able to demonstrate ingestion of *P. berghei*-infected erythrocytes by splenic macrophages activated by malarial infection and peritoneal macrophages activated with thioglycollate.[22]

On the basis of these findings, we can conclude that two factors enhance the ingestion of parasitized erythrocyte: macrophage activation and immune serum directed at the surface of the infected erythrocytes. Macrophage activation enhances ingestion although it is apparently not absolutely required;[52,53] however, immune serum or antibody sensitization does appear to be essential.[54] In addition to free parasites, mature erythrocytes appear to be the targets of ingestion, probably because they express the surface antigens against which the immune serum is directed (Section III).

What is the relevance of these results to protection of the host *in vivo*? This question has been most clearly addressed by the work of Fandeur et al.[58] and Michel et al.[59] These investigators immunized Saimiri monkeys with *P. falciparum* and purified the resulting immunoglobulins (Igs) from ascites fluid produced in the animals. These Igs were then assayed for their ability to either protect monkeys passively from *P. falciparum* or neutralize the ability of merozoites of this parasite to invade erythrocytes *in vitro*. Although sera with either protective activity *in vivo* or neutralizing activity were found, their activities were mutually exclusive, i.e., the protective sera did not neutralize merozoites and the neutralizing sera did not protect monkeys from subsequent challenge *in vivo*. Further, Fab'2 fragments of the protective sera were not effective, indicating that, *in vivo*, these sera probably interact with Fc receptor[+] cells in order to mediate protection. Indeed, these studies[59] showed that the protective ascites fluids mediated phagocytosis of *P. falciparum*-infected erythrocytes *in vitro*. Of interest in this regard is the study of Majarian et al.[60] in which a monoclonal antibody against *P. yoelii* 17x achieved protection by passive transfer. The protected animals still developed a patent parasitemia which was then cleared. In this instance too, the authors suggest the possibility that another mechanism, perhaps cellular, is involved in the clearance of the parasite.

It is worth noting that the protective antibodies described above were active at concentrations estimated to be as low as 0.1 mg/ml. In contrast, monoclonal antibodies which neutralize merozoites *in vitro* are generally not effective below 1 to 2 mg/ml[62,62] which would be difficult to achieve and maintain in the serum of an infected or immunized host. Therefore, antibodies against the erythrocytic stages of malaria are more likely to act together with effector cells which have Fc receptors for IgG.

B. THE ROLE OF THE SPLEEN AND IMMUNE CLEARANCE DURING MALARIA

The spleen plays a special role in immunity to malaria. In most cases, removal of the spleen results in the recurrence or relapse of an otherwise latent malaria infection.[63] Furthermore, species of malaria not normally infectious to a particular host will infect that host if it is splenectomized.[64] This role of the spleen in controlling malarial infections can most likely be attributed to macrophage activity. In man, the spleen contains about one seventh of the total reticuloendothelial tissue of the body. The spleen is the organ largely responsible for the removal of aged, deformed, and antibody-sensitized erythrocytes.[65] As we have seen, splenic macrophages are highly activated during malarial infections and are at least partially reponsible for keeping infection in check. In addition, it has been found that an architecturally intact spleen is necessary to control nonlethal infections since reconstitution of mice with spleen cell suspensions could not substitute for an intact spleen.[66]

Clearance studies have been done to determine the role of the spleen in clearing malaria infections and the nature of the interaction between the parasitized erythrocytes and splenic macrophages. The studies of Quinn and Wyler[67] in the rat indicated that immune rats have an enhanced capacity to clear erythrocytes infected with *P. berghei* and that the spleen was primarily responsible for this activity. They found that the rate of clearance of infected erythrocytes did not change over the course of infection until the crisis phase when clearance was markedly enhanced. In addition, they compared the clearance of normal erythrocytes, heavily sensitized with antierythrocyte IgG, with the clearance of erythrocytes containing Heinz bodies, in infected rats. They found the clearance of infected erythrocytes to resemble the clearance of rheologically altered red cells. In mice, Smith et al.[68] also found clearance of *P. yoelii*-infected erythrocytes to be spleen dependent but antibody independent.

In this laboratory, clearance studies were performed in mice infected with *P. berghei*.[69] We looked at the ability of infected mice to clear optimally sensitized erythrocytes, suboptimally sensitized erythrocytes, and complement-sensitized erythrocytes over the course of infection. Our previous observations that malaria-infected erythrocytes are coated with low levels of IgG[13] prompted us to use suboptimally sensitized erythrocytes as a model for infected erythrocytes. Our findings differed from those of Quinn and Wyler[67] in that clearance rates of optimally sensitized erythrocytes were greatly enhanced during the 1st week of infection. After 2 weeks of infection, however, the mice were no longer able to recognize antibody-coated red cells (Figure 2). We found this to be due to the profound complement depletion which occurs during malaria[70,71] since erythrocytes sensitized with complement-fixing levels of IgG and complement were cleared by mice at this stage of infection.[69]

In addition, we found that, during the early phase of the disease, malaria-infected mice could clear suboptimally sensitized erythrocytes which normal animals could not recognize.[69] This clearance was mediated by the spleen, not the liver, and we attribute this activity to the presence of highly activated macrophages which we observed *in vitro*.[22]

During the late phase of infection, the suboptimally sensitized erythrocytes were also not cleared. This was not due to depletion of complement since suboptimally sensitized erythrocytes do not fix complement. Rather, this defect was thought to be due to inhibition of macrophages by immune complexes. Serum taken from *P. berghei*-infected mice inhibited phagocytosis *in vitro*,[22,72] as did partially purified immune complexes.[73,74] More recently, we found that sera obtained from both *P. falciparum*-infected Saimiri monkeys and humans inhibited phagocytosis by human monocytes *in vitro*.[75] Interestingly, a study of the phagocytic ability of human monocytes of *P. falciparum*-infected patients revealed enhanced Fc receptor function in uncomplicated malaria but not in cells of patients with cerebral complications.[76] Whether this is due to a change in receptor number or function or to blockage of receptors by immune complexes is not known.

The differences in the studies of Quinn and Wyler[67] and ourselves are largely due to the differences in the models used. *P. berghei* in the rat is not lethal and the animals undergo a clear crisis phase. Indeed, the induction of crisis is highly dependent on the presence of the spleen.[77] In mice, *P. berghei* is lethal and the effect on the spleen is probably closer to that described by Weiss et al.[10] for lethal infection with *P. yoelii* 17x (Section I). Thus, we find an initial enhancement of clearance until the late phase of the disease when the macrophages of the host are overwhelmed.

Indeed, several studies indicate that macrophages may contribute to immunosuppression in malaria. Studies of Loose,[78] Weidanz,[49] and others indicate that splenic macrophages may be defective as accessory cells in *in vitro* antibody responses, perhaps due to the recruitment of IA⁻ macrophages to the spleen. In addition, Murphy[80] found that at the peak

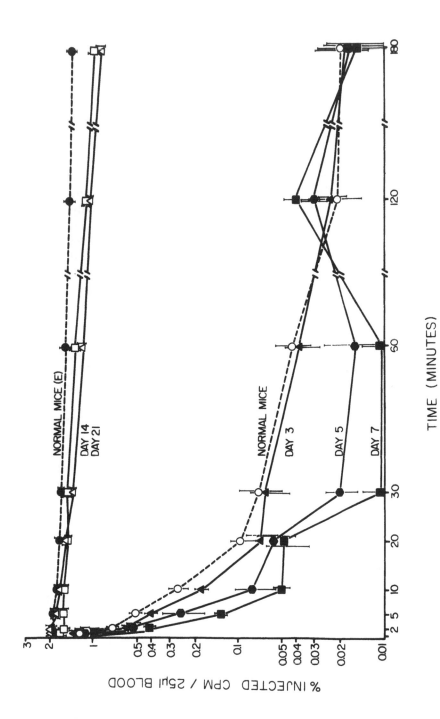

FIGURE 2. Clearance of ^{51}Cr-labeled 4×10^7 E or EIgG.10^{-3} (E sensitized with a 1:1000 dilution of antibody) in normal and *P. beghei*-infected mice. Clearance was determined from 25-μl blood samples taken at the time points indicated. (–●–) E injected into normal mice, (–○–) EIgG.10^{-3} injected into normal mice, and in mice 3 d (▲), 5 d (–●–), 7 d (–■–), 14 d (□), and 21 d (△) after infection with *P. berghei*. Data points represent the mean values ± 1 SEM of 6 to 14 mice in at least three experiments. (From *J. Clin. Invest.*, 67, 183, 1981. With permission.)

of *P. yoelii* 17x infection, macrophage microbicidal activity for *Listeria monocytogenes* was impaired. These data are consistant with our finding[67] that the ability of the monocyte-macrophage system to clear sensitized particles changes over the course of infection. The mechanisms of macrophage immunosuppression are as yet undefined but certainly the effect of immune complexes[72-74] and the recruitment of IA⁻ macrophages to the spleen probably both contribute to macrophage defects. Wyler and Gallin[20] also noted a change in the nature of the soluble mediators secreted by splenic adherent cells over the course of rodent malaria infections. In addition, in human malarial infections, chemotactic activity of monocytes and neutrophils is suppressed prior to treatment.[81]

C. KILLING OF INTRAERYTHROCYTIC MALARIA PARASITES

There are several known cellular killing mechanisms which could be directed against intraerythrocytic malarial parasites, namely, T cell-mediated cytotoxicity, antibody-dependent cell-mediated cytotoxicity (ADCC), and direct killing by macrophages, NK cells, neutrophils, or eosinophils. T cell-mediated cytotoxicity against intraerythrocytic malarial parasites has only been demonstrated once.[81] Since T dependent erythrocytes express only very low levels of H-2 molecules, this is an unlikely killing mechanism for parasites within mature erythrocytes. Therefore, killing of intraerythrocytic malaria parasites is probably not mediated by classic cytotoxic T cells. Direct killing of intraerythrocytic parasites by NK cells has not been demonstrated, although an association between NK activity and resistance to malaria has been suggested.[82] Phagocytosis by eosinophils[52] and, very recently, killing of *P. falciparum* by neutrophils[83] has been demonstrated.

This discussion will focus on macrophage-mediated killing. Early evidence that macrophages and/or monocytes might kill intraerythrocytic malaria parasites was provided by the studies of Criswell et al.[84] They implanted Millipore chambers containing *P. berghei*-infected erythrocytes and macrophages into the peritoneal cavities of mice. Macrophages obtained from mice with chronic infection or normal macrophages placed into chronically infected mice enhanced the clearance of parasitemia. *In vitro*, Langhorne et al.[85] showed that incubation of spleen cells from infected monkeys with parasitized erythrocytes reduced the ability of the parasites to multiply. After this, Taverne et al.[86] suggested that in mice *P. yoelii* was killed by cells of the monocyte-macrophage series. Infected erythrocytes were incubated with various populations of effector cells and then injected into mice. Parasite survival was determined by measuring the time after injection when parasitemia reached 0.5% and then calculating the number of parasites injected based on a standard curve. The most effective cell population was a phagocytic cell which was not a fully mature macrophage, i.e., insensitive to antimacrophage antiserum. Data from this study also showed that fresh blood monocytes or peritoneal cells activated by incubation with lymph node cells of immunized mice were more effective than normal peritoneal cells. While all these studies suggested that activated macrophages could kill malaria parasites within red cells, these experiments did not differentiate between cytotoxic effects of the macrophages and phagocytic effects.

In this laboratory, similar incubation and injection studies were done. These studies were then extended as described below and demonstrated killing of intraerythrocytic parasites by appropriately activated macrophages.

In our initial experiments,[87] macrophages were incubated in direct contact with *P. yoelii* 17x-parasitized erythrocytes. After 13 to 15 h, the incubation mixture was injected into mice. A sensitive radioimmunoassay for the parasite was used in order to determine parasitemia in the recipient mice 3 d after inoculation. We found that macrophages from either BCG or *P. yoelii*-infected mice were more effective in killing the intraerythrocytic parasites than normal macrophages. Further, macrophages, incubated with supernatants derived from the spleens of malaria-infected animals,[88] were highly inhibitory to the parasitized eryth-

TABLE 1
Effect of Hyperimmune Serum on the Killing of *P. yoelii* PE by LK-Activated Macrophages

Macrophage	Phagocytic stimulus[a]	Reciprocal antibody titer[b]	(^{3}H) isoleucine incorporation compared with control[c]
−	−	−	100
+	−	−	96 ± 6
+	+	−	65 ± 3
+	+	10^2	50 ± 2
+	+	10^3	41 ± 6
+	+	10^4	64 ± 4
+	PMA (200 ng/ml)	−	62 ± 4

Note: Resident macrophages (4.5 × 10^6) were stimulated for 48 h by a LK prepared from the spleens of *P. yoelii*-infected mice. Target PE (1 × 10^7) were suspended in Adaps chambers separated from macrophages by a 0.45-μm filter.

[a] Lysed PE (described in Materials and Methods) served as the phagocytic stimulus to macrophages in bottom chamber of culture vessel.
[b] Hyperimmune serum (final dilution 10^{-2} to 10^{-4}) was added with lysed PE to macrophage monolayers.
[c] Results expressed as the mean cpm (^{3}H) isoleucine incorporation ± SE of triplicate samples compared with control PE incubated without hyperimmune serum and in the absence of effector cells.

From Ockenhouse, C. F. and Shear, H. L., *J. Immunol.*, 132, 424, 1984. With permission.

rocytes. However, when macrophages were preincubated with cytochalasin B, a potent phagocytic inhibitor, killing of the parasites was only slightly reduced. This result suggested that phagocytosis was not solely responsible for the elimination of the parasites.

In order to determine whether macrophages could indeed kill intraerythrocytic parasites without ingestion, we devised a culture system to assay killing in which the target cells and effector cells were physically separated.[87] Effector macrophages, cultured in the bottom of a 24-well tissue culture plate, were stimulated *in vitro* with culture supernatants from the spleens of mice that were either BCG- or malaria-infected. The parasitized erythrocytes were suspended above the macrophages in Adaps chambers, in which the macrophages were separated from the erythrocytes by a Millipore filter with a pore size of 0.45 μm. After 18 h, the ability of the parasites to incorporate tritiated amino acids was determined as a measure of viability. As shown in Table 1, protein synthesis of target parasites was partially inhibited when macrophages were activated by BCG or *P. yoelii*-derived LK and not inhibited by resident macrophages. However, cytotoxicity could be greatly enhanced if a phagocytic stimulus (such as lysed parasitized erythrocytes) was added to the macrophage monolayer (Table 1). The addition of hyperimmune serum to the lysed parasitized erythrocytes further enhanced their capacity to act as a phagocytic trigger. Studies on the mechanism of this killing are described below (Section IV.D).

Further studies from this laboratory indicate that killing of intraerythrocytic malaria parasites may also occur in the human infection *P. falciparum*.[89] *In vitro*, monocytes enriched from human peripheral blood leukocytes inhibited the multiplication of *P. falciparum* by 63%, compared to parasites grown in the absence of effector cells. Although the erythrocytes were not physically separated from the monocytes as in the study discussed above,[87] phagocytosis of parasitized erythrocytes was not observed. Indeed, binding of parasitized erythrocytes to monocytes was sufficient to stimulate killing. Subsequent studies have shown that *P. falciparum* binds to monocytes via their knobs[90] and this is consistent with our observation that ring stage-parasitized erythrocytes did not stimulate an oxidative response in the monocytes.

Another potential mechanism of macrophage-mediated killing of intraerythrocytic parasites is antibody-dependent cell-mediated cytotoxicity. ADCC is an effector mechanism which operates against tumor cells and other target cells sensitized with specific antibodies against surface membrane antigens and is mediated by a subpopulation of heterogeneous lymphocytes called K cells as well as mononuclear cells and granulocytes.[91] As has been discussed (Section III), several plasmodial antigens are expressed on or in the erythrocyte membranes of malaria-infected erythrocytes and could be the target(s) of ADCC.

ADCC has been observed when mononuclear cells and malaria-immune serum are incubated with *P. falciparum*-infected erythrocytes in culture.[92] However, a ratio of 50 lymphocytes per parasitized erythrocyte was necessary to demonstrate cytotoxicity. Other studies measured ADCC against unrelated targets during malaria. Gilbreath et al.[93] found ADCC depressed during human malaria infections using Chang cells as targets and unchanged using antibody-coated erythrocytic targets. In contrast, ADCC against antibody-coated chicken erythrocytes was enhanced in mice infected with *P. chabaudi*.[94] Thus, the data to date indicates that the degree of ADCC during malaria may vary in different clinical and experimental situations and with malaria-infected erythrocytes or unrelated targets.[92-96]

Our own studies[131] were initiated to clarify some of these observations. We used several rodent models of infection and looked at the ability of splenic cells to lyse either autologous mouse erythrocytes sensitized with antierythrocyte IgG or malaria-infected erythrocytes, with or without the help of immune serum.

Early in infection, spleen cells from mice infected with *P. berghei* had enhanced cytotoxicity against normal mouse erythorcytes sensitized with IgG. However, later in infection cytotoxicity was suppressed. Serum and partially purified immune complexes, taken from *P. berghei*-infected mice late in infection, inhibited ADCC. In addition, spleen cells from mice infected with the lethal variant of *Plasmodium* 17X were less cytotoxic than cells from mice infected with the nonlethal variant. Thus, there seem to be at least two levels of control on ADCC during rodent malaria. One, exemplified in the *P. berghei* experiment, may be regulated by immune complexes. The other control mechanism appears to be associated with the virulence of the parasite but we do not know what the nature of this control is.

Curiously, we did not observe ADCC by activated macrophages against malaria-infected erythrocytes sensitized with hyperimmune serum. In contrast, such macrophages readily lysed infected erythrocytes sensitized with antierythrocyte antibodies. This result might be due to the lack of antibody of the appropriate specificity in the serum. It is also possible that parasite antigens of the erythrocytes are continuously shed. As discussed above, some erythrocyte membrane parasite antigens are only visualized after the erythrocytes are fixed with gluteraldehyde. Further studies should clarify this point.

D. MACROPHAGE-DERIVED MEDIATORS OF CYTOTOXICITY

Early evidence of a toxic role for macrophage products on the asexual stages of malaria was the finding of Clark et al.[97] that pretreatment of mice with *Mycobacterium bovis*, strain BCG, 1 month prior to infection with *P. berghei* or *P. vinckei*, partially protected the animals. Further, the blood films of the protected animals showed evidence of intraerythrocytic death of the parasites. This was followed by studies showing that the injection of mice with compounds which generate reactive oxygen intermediates *in vivo*; alloxan[98] and *t*-butyl hydroperoxide[99] reduce parasitemia of *P. vinckei*-infected mice. Similarly, Taverne, et al. found that injection of H_2O_2 directly into mice partially protected them from *P. yoelii* infection.[100] Furthermore, the direct cytotoxic effects of H_2O_2 on *P. yoelii*,[100] *P. berghei*,[100] and *P. falciparum*[89,101] *in vitro* has been observed.

Support for the idea that the toxic oxygen metabolites came from macrophages were the findings that parasitized erythrocytes could elicit a chemiluminescence response in normal mouse peritoneal macrophages which was enhanced by immune serum[102] and that splenic macrophages from *P. yoelii*-infected mice (a nonlethal infection) generated larger amounts of O_2- and H_2O_2[103,104] than did macrophages obtained from *P. berghei*-infected animals[104] (a lethal infection). Interestingly, recent data[105] indicates that merozoites and antigens derived from culture supernatants of *P. falciparum* are able to elicit a chemiluminescence response from normal human neutrophils and monocytes, suggesting that the radicals so generated could be involved in both protective and pathogenic effects, as previously suggested.[106]

Finally, the prevalence of B-thalassemia and glucose-6-phosphate dehydrogenase (G6PD) deficiency in malarious areas of the world are thought to be due to the enhanced resistance provided by these traits against malaria. Friedman[107] has shown that *P. falciparum* grown in B-thalassemia trait erythrocytes are more sensitive to oxidative stress and that the parasites could be protected by antioxidants such as vitamin E. Roth, et al.[108] found that *P. falciparum* grew less efficiently in G6Pd-deficient erythrocytes, although the biochemical basis for this was not defined. Thus, the basis for maintenance of these traits might be their enhanced sensitivity to the release of reactive oxygen metabolites.

Recently, the effect of B-thalassemia in mice with homozygous B-thalassemia syndrome was studied.[109] These mice were partially protected from *P. chabaudi* infection (which invades mature erythrocytes) whereas *P. berghei* infection (which preferentially invades reticulocytes) was slightly more severe. The enhanced *P. berghei* infection was most likely due to the anemia and high reticulocyte count in these mice. Interestingly, splenectomized, B-thalassemic mice infected with *P. chabaudi*, had even more delayed parasitemia. Although the finding in splenectomized mice is as yet unexplained, this model should be useful in elucidating the protection of this hemoglobinopathy.

In our cytotoxicity studies described above (Section IV.C), we wanted to determine whether reactive oxygen intermediates were responsible for the inhibition of leucine incorporation by *P.yoelii 17x*[87] and *P. falciparum*.[89] Therefore, we cultured stimulated effector cells in the presence of several scavengers of reactive oxygen intermediates to determine whether cytotoxicity would be affected.[87] As shown in Table 2, catalase, a scavenger of H_2O_2, partially inhibited killing of *P. yoelii* whereas autoclaved catalase had no effect. In addition, macrophages that were pretriggered and then cultured in glucose-free medium, and therefore unable to produce H_2O_2, also were inhibited from killing. Other scavengers such as superoxide dismutase (SOD) histidine, and mannitol were without effect. However, these experiments do not exclude the possibility that other intermediates also damage the intraerythrocytic parasites.

These results not only demonstrate killing of intraerythrocytic malaria parasites by activated macrophages, they also suggest that phagocytosis of parasitized erythrocytes, especially after the parasites are sensitized by immune serum, is an integral part of the triggering of the oxidative burst necessary for killing. Indeed, release of H_2O_2 by granulocytes subsequent to phagocytosis has been demonstrated previously.[110]

As we observed in the rodent studies,[87] H_2O_2 appears to be the mediator of killing of *P. falciparum* based on the following evidence.[89] The degree of cytotoxicity could be correlated with the level of H_2O_2 released and killing could be inhibited by including catalase and not SOD in the incubation mixture. However, the possibility that other reactive intermediates also participate in damage to *P. falciparum* has not been excluded.

Monocytes which were cultured *in vitro* for 72 h or more (monocyte-derived macrophages) were no longer cytotoxic to *P. falciparum*. We, therefore, attempted to restore their activity with a T cell-derived LK and recombinant human γ-interferon.[89] Both concanavalin A-induced LK-, as well as γ-interferon-activated macrophages inhibited the multiplication of *P. falciparum* by approximately 60%. A titration of the dose of γ-interferon needed to activate macrophages for cytoxicity revealed that as little as 1 U/ml was sufficient to activate macrophages to inhibit the growth of *P. falciparum*. In addition, there was a positive correlation between the amount of interferon used, the degree of parasite inhibition, and the

TABLE 2

Participation of Reactive Oxygen Intermediates in the Killing of *P. yoelii* PE by LK-Activated Macrophages

Phagocytic stimulus[a]	Scavenger	(^{3}H) Isoleucine incorporation[b] (%)
None	None	91.0 ± 6
PMA (500 ng/ml)	None	40.0 ± 7
Zymosan (0.5 mg/ml)	None	33.0 ± 9
PE (2 × 10^7)	None	31.0 ± 4
PE	Catalase, 500 μg/ml	66.0 ± 7
PE	Catalase, inactivated	39.0 ± 5
PE	SOD, 1 mg/ml	41.0 ± 4
PE	Histidine, 10 mM	23.0 ± 6
PE	Mannitol, 50 mM	37.0 ± 7
PE	Pretriggered with PMA; glucose-depleted for 2 h	86.0 ± 11

Note: Resident macrophages (4 × 10^6) were stimulated for 48 h by a LK prepared from spleens of *P. yoelii*-infected mice. Target PE (1 × 10^7) were suspended in Adaps chambers separated from macrophages by a 0.45-μm filter.

[a] Phagocytic stimulus was added to the macrophage monolayer incubated in the bottom chamber of the culture vessel. Only the target PE in the upper chamber were assayed for (^{3}H) isoleucine incorporation.
[b] Results expressed as the mean cpm (^{3}H) isoleucine incorporation ± SE of triplicate samples compared with control PE incubated in the absence of effector cells.

From Ockenhouse, C. F. and Shear, H. L., *J. Immunol.*, 132, 424, 1984. With permission.

amount of H$_2$O$_2$ generated (Figure 3). Furthermore, those parasites which survived exposure to the activated macrophages developed more slowly than did the parasites in control cultures and many of the intraerythrocytic parasites appeared to be degenerating, as evidenced by their pyknotic nuclei, characteristic of the crisis forms described by Taliaferro and Cannon[7,8] (Figure 4).

Monocytes obtained from a patient with chronic granulomatous disease (CGD) of childhood, which fail to respond oxidatively to phagocytic stimuli, did not inhibit parasite growth. However, after lymphokine stimulation, CGD cells were somewhat inhibitory, suggesting that oxygen-independent mechanisms of inhibition also occur.[89]

In recent studies, we have utilized recombinant mouse γ-interferon to explore the role of this lymphokine in resistance to malaria. SW female mice, were injected with doses of γ-interferon ranging from 1000 to 100,000 U/d. We found that this treatment protected mice, in a dose-dependent fashion, from the lethal variant of *P. yoelii* 17X but not the nonlethal strain.[111] In addition, resistance to infection could be correlated with the ability of mouse splenocytes of several strains to produce both γ-interferon and H$_2$O$_2$ (Shear, H. L., in preparation). We hypothesize that the role of γ-interferon is to activate macrophages for killing intraerythrocytic stages of malaria, as the *in vitro* studies suggest.[89] The studies of Clark et al.[112] indicate that treatment of mice with recombinant γ-interferon protects them against *P. chabaudi*, indicating that this may be a general phenomenon in immune resistance to blood stage parasites.

There are other circulating factors which may damage malaria parasites within red cells. One of the earliest of those described was a factor induced by injection of a small amount endotoxin into either BCG, *Corynebacterium parvum* or malaria-infected mice[113,114] called tumor necrosis serum because it also was cytotoxic to tumor cells. Tumor necrosis serum was found to cause the death of malaria parasites *in vivo*[115] and *in vitro*.[116,117] This factor was shown to be secreted by macrophages of LPS-responsive[118] mice but not by macrophages

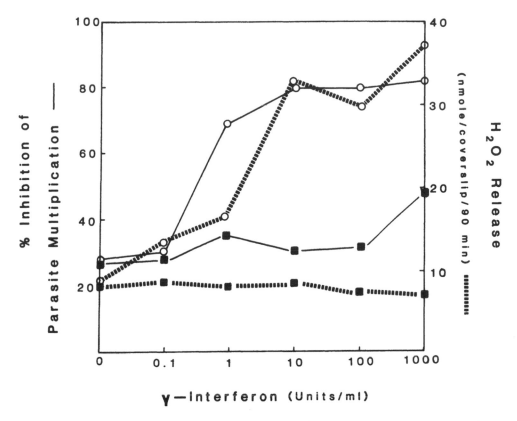

FIGURE 3. Correlation between parasite killing and H_2O_2 release from monocyte derived macrophages (MDM) activated with increasing concentrations of IFN-γ. MDM were incubated for 72 h before the addition of parasitized erythrocytes with IFN-γ (○) or IFN-γ neutralized by monoclonal anti-IFN-γ antibody (■), inhibition of multiplication (——) was assayed as described in Materials and Methods, and the results are compared to cultures incubated in the absence of effector cells. Concurrently, H_2O_2 release by activated MDM stimulated with phorbol myristate acetate (PMA) (500 ng/ml) was measured (----). The results represent the mean of four cultures + / − SE. (From Ockenhouse, C. F., Schulman, S., and Shear, H. L., *J. Immunol.*, 133(3), 1601, 1984. With permission.)

of LPS-nonresponder mice.[119] Recently, Clark et al.[112] found that human recombinant TNF released from osmotic pumps had an inhibitory effect on *P. chabaudi* infection in mice.

"Crisis forming factor", which damages the erythrocytic stages of *P. falciparum*, is discussed in Chapter 5.

V. THE ROLE OF MACROPHAGES IN VACCINATION AND PROPHYLAXIS AGAINST MALARIA

As suggested in Section I, the goal of understanding the role of macrophages in the immune response to malaria is to utilize this information in enhancing or developing immuno- or chemoprophylaxis against malaria. There is some experimental work suggesting that this may be possible.

Early evidence for a role for cell-mediated immunity in vaccination against the erythrocytic stages of malaria came from Playfair's laboratory.[120,121] These investigators observed that, in mice immunized against the rodent malarias, *P. berghei* and *P. yoelii*, the most effective vaccine included the addition of *Bordetella pertussis*. Protection was associated with both antibody responses and delayed type hypersensitivity, mediated by specifically sensitized T cells. They suggested that the function of the T cells might be to produce the

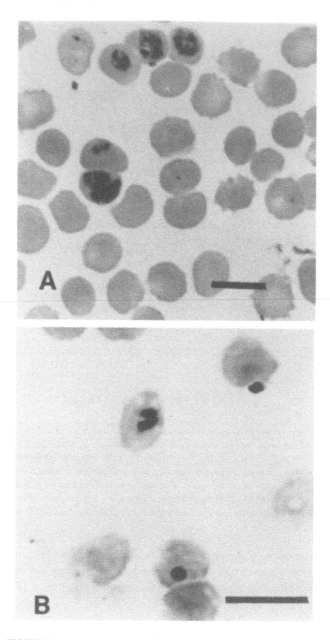

FIGURE 4.   Induction of crisis forms in parasitized erythrocytes incubated with IFN-γ-activated monocyte-derived macrophages. Erythrocytes cultured in the absence of effector macrophages contain well-developed, viable trophozoites (A), while intraerythrocytic parasites exposed to effector cells appear degenerated with pyknotic nuclei (B). Bar = 10 μm. (From Ockenhouse, C. F., Schulman, S., and Shear, H. L., *J. Immunol.*, 133(3), 1601, 1984. With permission.)

appropriate lymphokine(s) that would recruit and activate the necessary effector cells. Vaccination, then, would bring together the parasites, antibodies, and effector cells. In a more recent study, this group examined the appearance of cytotoxic spleen cells during lethal and nonlethal malarial infections of mice and in vaccinated animals.[122] They observed that cytotoxic cells appeared in both lethal and nonlethal infections but that nonlethal infection gave rise to a second peak of cytotoxic cells at the time of resolution of the infection.

Furthermore, vaccination led to an earlier and increased cytotoxic response which could be passively transferred with lymphocytes. Again, the possibility that lymphokines produced by the transferred cells enhanced cytotoxicity was suggested.

Additional evidence implicating a cell-mediated component in resistance to challenge comes from vaccination studies with either purified antigens or peptides. One of the first of these studies was that of Freeman and Holder.[123] They immunized mice with a purified antigen of *P. yoelii*, the 230-kDa antigen, in saponin, and challenged with the homologous parasite. Although the animals were partially protected, they did develop a patent parasitemia for 5 d which was then cleared. Crisis forms were found in the blood smears of these mice. Immune serum of the immunized animals did not transfer protection, although serum from the challenged mice did. The authors suggest that cell-mediated immunity (CMI) may have played a role in the clearance of the parasite.

Indeed, in all instances in which protection is achieved against the erythrocytic phase of infection, challenge produces parasitemia, usually lower than in unvaccinated animals, which is then cleared, (e.g., see References 44, 47, and 123 to 125). The nature of the immune mechanism that is induced, which seems to act synergistically with the vaccine, is as yet unidentified, however it appears to be cellular rather than humoral.

Based on the data reviewed in this chapter on the potent effects of macrophages on intraerythrocytic parasites and the dependence on T cells for the activation of macrophages, it would appear that efforts to trigger T cells for macrophage activation would be a reasonable vaccination strategy. Indeed, the importance of CMI in protection against helminths[126] and the sporozoite stage of malaria[127,128] has only recently been emphasized.[129] Recently, however, Playfair and De Souza[130] utilized γ-interferon as an adjuvant in vaccination against *P. yoelii* YM (a lethal infection) and found it to be almost as effective as saponin in protecting the mice against challenge. Mice that received either saponin or γ-interferon had both enhanced humoral as well as CMI. Further studies are needed to elucidate the mechanism of action of the γ-interferon.

Our knowledge of how macrophages may destroy the asexual stages of Plasmodium may also be useful in the design of new drugs for malaria. Studies by Clark and colleagues discussed above have shown that injection of experimental animals with alloxan and *t*-butyl hydroperoxide was effective in reducing parasitemia. In developing drugs of this type, consideration of the systemic effects of this type of chemotherapy will obviously be important.

VI. CONCLUSIONS

Macrophages appear to be an important component in the immune response to malaria. Phagocytosis of asexual parasites and parasitized erythrocytes, observed early in the study of this infection, is greatly enhanced by macrophage activation and by antibodies directed against the surface of parasitized red blood cells. Antibodies against this stage of the life cycle of malaria appear to function cooperatively with a cellular component of the immune response, likely to include macrophages and perhaps other Fc receptor[+] cells. Killing of intraerythrocytic malaria parasites is also mediated by macrophages via the secretion of H_2O_2 and perhaps other toxic substances. Such killing is enhanced by activation of the macrophages by lymphokines such as γ-interferon and by phagocytosis itself. Further attempts to enhance macrophage antiplasmodial immunity must involve the identification of antigens and adjuvants which trigger either T cells to produce the appropriate macrophage stimulants or directly influence macrophage function. This information may also help in the design of drugs which function similarly to destroy asexual malaria parasites, without harming the host.

ACKNOWLEDGMENTS

This work received support from the National Institutes of Health grant AI15235; the

U.S. Army, contract DAMD17-85-C-5175, and the World Health Organization, contract T16/181/M2/1. I am also grateful to Dr. Elizabeth Nardin for reviewing the manuscript and Ms. Luna Ming for secretarial assistance.

REFERENCES

1. **Unanue, E. R. and Allen, P. M.,** The basis for the immunoregulatory role of macrophages and other accessory cells, *Science*, 236, 551, 1987.
2. **Nathan, C. F.,** Secretory products of macrophages, *J. Clin. Invest.*, 79, 319,1987.
3. **Evans, R. and Alexander, P.,** Cooperation of immune lymphoid cells with macrophages in tumour immunity, *Nature*, 228, 620, 1970.
4. **Murray, H. W.,** Cellular resistance to protozoal infection, *Annu. Rev. Med.*, 37, 61, 1986.
5. **Nogueira, N., Gordon, S., and Cohn, Z.,** *Trypanosoma cruzi:* modification of macrophage function during infection, *J. Exp. Med.*, 146, 157, 1977.
6. **Taliaferro, W. H. and Mulligan, H. W.,** The histopathology of malaria with special reference to the function and origin of the macrophages in defence, *Indian Med. Res. Mem.*, 29, 1, 1937.
7. **Cannon, P. R. and Taliaferro, W. H.,** Acquired immunity in avian malaria. III. Cellular reactions in infection and superinfection, *J. Prev. Med.*, 5, 37, 1931.
8. **Taliaferro, W. H. and Cannon, P. R.,** The cellular reactions during primary infections and superinfections of *Plasmodium brasilianum* in Panamanian monkeys, *J. Inf. Dis.*, 59, 72, 1936.
9. **Taliaferrro, W. H. and Taliaferro, L. G.,** The effect on immunity on the asexual reproduction of *Plasmodium brasilianum, J. Infect. Dis.*, 75, 1, 1944.
10. **Weiss, L., Geduldig, U., and Weidanz, W.,** Mechanisms of splenic control of murine malaria: reticular cell activation and the development of a blood-spleen barrier, *Am. J. Anat.*, 176, 251, 1986.
11. **Zuckerman, A.,** Recent studies on factors involved in malarial anemia, *Mil. Med.* 1201, 1966.
12. **Zuckerman, A., Spira, D. T., and Ron, N.,** A quantitative study of phagocytosis in the spleen of rats infected with *Plasmodium berghei, Dynamic Aspects of Host-Parasite Relationships*, Vol. 1, Academic Press, New York, 1973, 79.
13. **Lustig, H. J., Nussenzweig, V., and Nussenzweig, R. S.,** Erythrocyte membrane-associated immunoglobulins during malaria infection of mice, *J. Immunol.*, 119, 210, 1977.
14. **Hunter, K. W., Finkelman, F. D., Strickland, G. T., Sayles, P. C., and Scher, I.,** Murine malaria: analysis of erythrocyte surface bound immunoglobulin by flow microfluorimetry, *J. Immunol.*, 125, 169, 1980.
15. **Facer, C. A., Bray, R. S., and Brown, J.,** Direct Coombs antiglobulin reactions in Gambian children with *Plasmodium falciparum* malaria. I. Incidence and class specificity, *Clin. Exp. Immunol.*, 35, 119, 1979.
16. **Facer, C. A.,** Direct Coombs antiglobulin reactions in Gambian children with *Plasmodium falciparum* malaria. II. Specificity of erythrocyte bound IgG, *Clin. Exp. Immunol.*, 39, 279, 1980.
17. **Facer, C. A.,** Direct antiglobulin reactions in Gambian children with *Plasmodium falciparum* malaria. III. Expression of IgG subclass determinants and genetic markers and association with anaemia, *Clin. Exp. Immunol.*, 41, 81, 1980.
18. **Abdalla, S. and Weatherall, D. J.,** The direct antiglobulin test in *P. falciparum* malaria, *Br. J. of Haematol.*, 51, 415, 1982.
19. **Wernsdorfer, W. H.,** The importance of malaria in the world, in *Malaria*, Vol. 1, Kreier, J. P., Ed., Academic Press, New York, 1980, 47.
20. **Wyler, D. J. and Gallin, J. I.,** Spleen-derived mononuclear cell chemotactic factor in malaria infections: a possible mechanism for splenic macrophage accumulation, *J. Immunol.*, 118, 478, 1977.
21. **Lucia, H. L. and Nussenzweig, R. S.,** *Plasmodium chabaudi* and *Plasmodium vinckei:* phagocytic activity of mouse reticuloendothelial system, *Exp. Parasitol.*, 25, 319, 1969.
22. **Shear, H. L., Nussenzweig, R. S., and Bianco, C.,** Immune phagocytosis in murine malaria, *J. Exp. Med.*, 149, 1288, 1979.
23. **Lee, S., Crocker, P., and Gordon, S.,** Macrophage plasma membrane and secretory properties in murine malaria, *J. Exp. Med.*, 163, 54, 1986.
24. **Rudzinska, M. A. and Trager, W.,** The fine structure of trophozoites and gametocytes in *Plasmodium coatneyi, J. Protozool.*, 15, 73, 1968.
25. **Luse, S. A. and Miller, L. H.,** *Plasmodium falciparum* malaria, *Am. J. Trop. Med. Hyg.*, 20, 655, 1971.
26. **Miller, L. H.,** The ultrastructure of red cells infected by *Plasmodium falciparum* in man, *Trans. R. Soc. Trop. Med. Hyg.*, 66, 459, 1972.

27. **Langreth, S. G. and Reese, R. T.**, Antigenicity of the infected-erythrocyte and merozoite surfaces in falciparum malaria, *J. Exp. Med.*, 150, 1241, 1979.

28. **Kilejian, A.**, Characterization on a protein correlated with the production of knob-like protrusions on membranes of erythrocytes infected with *Plasmodium falciparum*, *J. Exp. Med.*, 151, 1534, 1979.

29. **Leech, J. H., Barnwell, J. W., Aikawa, M., Miller, L. H., and Howard, R. J.**, *Plasmodium falciparum* malaria: association of knobs on the surface of infected erythrocytes with a histidine-rich protein and the erythrocyte skeleton, *J. Cell Biol.*, 98, 1256, 1984.

30. **Udeinya, I., Schmidt, J. A., Aikawa, M., Miller, L. H., and Green, I.**, Falciparum malaria-infected erythrocytes specifically bind to cultured human endothelial cells, *Science*, 213, 55, 1981.

31. **Schmidt, J. A., Udeinya, I. J., Leech, J. H., Hay, R. J., Aikawa, M., Barnwell, J. W., Green, I., and Miller, L. H.**, *Plasmodium falciparum* malaria: an amelanotic melanoma cell line bears receptors for the knob ligand on infected erythrocytes, *J. Clin. Invest.*, 70, 379, 1982.

32. **Udeinya, I. J., Miller, L. H., McGregor, I. A., and Jensen, J. B.**, *Plasmodium falciparum* strain-specific antibody blocks binding of infected erythrocytes to amelanotic melanoma cells, *Nature*, 303, 429, 1983.

33. **Coppel, R. L, Culvenor, J. G., Bianco, A. E., Crewther, P. E., Stahl, H. D., Brown, G. V., Anders, R. F., and Kemp, D. J.**, Variable antigen associated with the surface of erythrocytes infected with mature stages of *Plasmodium falciparum*, *Mol. Biochem. Parasitol.*, 20, 265, 1986.

34. **Marsh, K. and Howard, R. J.**, Antigens induced on erythrocytes by *P. falciparum*: expression of diverse and conserved determinants, *Science*, 231, 150, 1986.

35. **Barnwell, J. W., Howard, R. J., and Miller, L. H.**, Influence of the spleen on the expression of surface antigens on parasitized erythrocytes, *Ciba Found. Symp.*, 94, 117, 1982.

36. **Hommel, M., David, P. H., and Oligino, L. D.**, Surface alterations of erythrocytes in *Plasmodium falciparum* malaria, *J. Exp. Med.*, 157, 1137, 1983.

37. **Barnwell, J. W., Howard, R. J., and Miller, L. H.**, Altered expression of *Plasmodium knowlesi* variant antigen on the erythrocyte membrane in splenectomized *Rhesus* monkeys, *J. Immunol.*, 128, 224, 1982.

38. **Barnwell, J. W., Howard, R. J., Coon, H. F., and Miller, L. H.**, Splenic requirement for antigenic variation and expression of the variant antigen on the erythrocyte membrane in cloned *Plasmodium knowlesi* malaria, *Infect. Immun.*, 40, 985, 1983.

39. **David, P. H., Hommel, M., Miller, L. H., Udeinya, I. J., and Oligino, L. D.**, Parasite sequestration in *Plasmodium falciparum* malaria: spleen and antibody modulation of cytoadherence of infected erythrocytes, *Proc. Natl. Acad. Sci. U.S.A.*, 80, 5075, 1983.

40. **Langreth, S. G. and Peterson, E.**, Pathogenicity stability, and immunogenicity of a knobless clone of *Plasmodium falciparum* in Colombian owl monkeys, *Infect. Immun.*, 47, 760, 1985.

41. **Perlmann, H., Berzins, K., Wahlgren, M., Carlsson, J., Bjorkman, A., Patarroyo, M. E., and Perlmann, P.**, Antibodies in malarial sera to parasite antigens in the membrane of erythrocytes infected with early asexual stages of *Plasmodium falciparum*, *J. Exp. Med.*, 159, 1686, 1984.

42. **Coppel, R. L., Cowman, A. F., Anders, R. F., Bianco, A. E., Saint, R. B., Lingelback, K. R., Kemp, D. J., and Brown, G. V.**, Immune sera recognize on erythrocytes a *Plasmodium falciparum* antigen composed of repeated amino acid sequences, *Nature*, 310, 789, 1984.

43. **Wahlin, B., Wahlgren, M., Perlmann, H., Berzins, K., Bjorkman, A., Patarroyo, M. E., and Perlmann, P.**, Human antibodies to a M_r 155,000 *Plasmodium falciparum* antigen efficiently inhibit merozoite invasion, *Proc. Natl. Acad. Sci. U.S.A.*, 81, 7912, 1984.

44. **Collins, W. E., Anders, R. F., Pappaioanou, M., Campbell, G. H., Brown, G. V., Kemp, D. J., Coppel, R. L., Skinner, J. C., Andrysiak, P. M., Favaloro, J. M., Corcoran, L. M., Broderson, J. R., Mitchell, G. F., and Campbell, C. C.**, Immunization of *Aotus* monkeys with recombinant proteins of an erythrocyte surface antigen of *Plasmodium falciparum*, *Nature*, 232, 259, 1986.

45. **Wanidworanun, C., Barnwell, J. W., Masuda, A., and Shear, H., L.**, Malarial antigen in the erythrocyte membrane of *Plasmodium chabaudi*-parasitized red blood cells, in *Molecular Strategies of Parasite Invasion*, Agabian, N., Goodman, H., and Nogueira, N., Eds., Alan R. Liss, New York, 1987, 355.

46. **Gabriel, J. A., Holmquist, G., Perlmann, H., Berzins, K., Wigzell, H., and Perlmann, P.**, Identification of a *Plasmodium chabaudi* antigen present in the membrane of ring stage infected erythrocytes, *Mol. Biochem. Parasitol.*, 20, 67, 1986.

47. **Wanidworanun, C., Barnwell, J. W., and Shear, H. L.**, Protective antigen in the membranes of mouse erythrocytes infected with *Plasmodium chabaudi*, *Mol. Biochem. Parasitol.*, 25, 195, 1987.

48. **Goldring, J., Burns, J., Long, C., Vaidya, A., Weidanz, W., Wanidworanun, C., and Shear, H. L.**, Partial characterization and recombinant expression of a protective antigen from the erythrocytic stage of *Plasmodium chabaudi adami*, *FASEB J.*, 2, A1257, 1988.

49. **Rodriguez, M. H. and Jungery, M.**, A protein on Plasmodium-infected erythrocytes functions as a transferrin receptor, *Nature*, 324, 388, 1986.

50. **Haldar, K., Henderson, C. L., and Cross, G. A. M.**, Identification of the parasite transferrin receptor of *Plasmodium falciparum*-infected erythrocytes and its acylation via 1,2-diacyl-sn-glycerol, *Proc. Natl. Acad. Sci.*, 83, 8565, 1986.

51. **Brown, K. N.,** Protective immunity to malaria provides a model for the survival of cells in an immunologically hostile environment, *Nature,* 230, 163, 1971.

52. **Tosta, C. E. and Wedderburn, N.,** Immune phagocytosis of *Plasmodium yoelii*-infected erythrocytes by macrophages and eosinophils, *Clin. Exp. Immunol.,* 42, 114, 1980.

53. **Celada, A., Cruchaud, A., and Perrin, L. H.,** Opsonic activity of human immune serum on *in vitro* phagocytosis of *Plasmodium falciparum*-infected red blood cells by monocytes, *Clin. Exp. Immunol.,* 47, 635, 1982.

54. **Khusmith, S., Druilhe, P., and Gentilini, M.,** Enhanced *Plasmodium falciparum* merozoite phagocytosis by monocytes from immune individuals, *Infect. Immun.,* 35, 874, 1982.

55. **Chow, J. S. and Kreier, J. P.,** *Plasmodium berghei:* adherence and phagocytosis by rat macrophages *in vitro, Exp. Parasitol.,* 31, 13, 1972.

56. **Brooks, C. and Kreier, J.,** Role of the surface coat in *in vitro* attachment and phagocytosis of *Plasmodium berghei* by peritoneal macrophages, *Infect. Immun.,* 20, 827, 1978.

57. **Brown, K. M. and Kreier, J.,** Effect of macrophage activation on phagocyte-plasmodium interaction, *Infect. Immun.,* 51, 744, 1986.

58. **Fandeur, T., Dubois, P., Gysin, J., Dedet, J. P., and Da Silva, L. P.,** *In vitro* and *in vivo* studies on protective and inhibitory antibodies against *Plasmodium falciparum* in the *Saimiri* monkey, *J. Immunol.,* 132, 432, 1984.

59. **Michel, J. C., Fandeur, T., Neuilly, G., Roussilhon, C., and Dedet, J. P.,** Opsonic activity of ascitic fluids from *Plasmodium falciparum*-infected *Saimiri* monkey: positive correlation with protection transfer assay, *Annu. Immunol.,* 134D, 373, 1983.

60. **Majarian, W. R., Daly, T. M., Weidanz, W. P., and Long, C. A.,** Passive immunization against murine malaria with an IgG3 monoclonal antibody, *J. Immunol.,* 132, 3131, 1984.

61. **Deans, J. A., Alderson, T., Thomas, A. W. G. H., Lennox, E. S., and Cohen, S.,** Rat monoclonal antibodies which inhibit the *in vitro* multiplication of *Plasmodium knowlesi, Clin. Exp. Immunol.,* 49, 297, 1982.

62. **Miller, L. H., Hudson, D., Rener, J., Taylor, D., Hadley, T. J., and Zilberstein, D.,** A monoclonal antibody to *Rhesus* erythrocyte band 3 inhibits invasion by malaria (*Plasmodium knowlesi*) merozoites, *J. Clin. Invest.,* 72, 1357, 1983.

63. **Garnham, P. C. C.,** The role of the spleen in protozoal infections with special reference to splenectomy, *Acta Trop.,* 27, 1, 1970.

64. **Todorovic, R., Ferris, D., and Ristic, M.,** Roles of the spleen in acute plasmodium and babesial infections in rats, *Exp. Parasitol.,* 21, 354, 1967.

65. **Frank, M. M., Schreiber, A. D., Atkinson, J. P., and Jaffe, C. J.,** Pathophysiology of immune hemolytic anemia, *Ann. Intern. Med.,* 87, 210, 1977.

66. **Oster, C. N., Koontz, L. C., and Wyler, D. J.,** Malaria in asplenic mice: effects of splenectomy, congenital asplenia, and splenic reconstitution on the course of infection, *Am. Soc. Trop. Med. Hyg.,* 29, 1138, 1980.

67. **Quinn, T. C. and Wyler, D. J.,** Intravascular clearance of parasitized erythrocytes in rodent malaria, *J. Clin. Invest.,* 63, 1187, 1979.

68. **Smith, L. P., Hunter, K. W., Oldfield, E. C., and Strickland, G. T.,** Murine malaria: blood clearance and organ sequestration of *Plasmodium yoelii*-infected erythrocytes, *Infect. Immun.,* 38, 162, 1982.

69. **Pappas, M. G., Nussenzweig, R. S., Nussenzweig, V., and Shear, H. L.,** Complement-mediated defect in clearance and sequestration of sensitized, autologous erythrocytes in rodent malaria, *J. Clin. Invest.,* 67, 183, 1981.

70. **Neva, F. A., Howard, W. A., Glew, R. H., Krotoski, W. A., Gam, A. A., Collins, W. E., Atkinson, J. P., and Frank, M. M.,** Relationship of serum complement levels to events of the malarial paroxysm, *J. Clin Invest.,* 54, 451, 1974.

71. **Krettli, A. U., Nussenzweig, V., and Nussenzweig, R. S.,** Complement alterations in rodent malaria, *Am. J. Trop. Med. Hgy.,* 25, 34, 1976.

72. **Packer, B. J. and Kreier, J. P.,** *Plasmodium berghei* malaria: effects of acute-phase serum and erythrocyte-bound immunoglobulins on erythrophagocytosis by rat peritoneal macrophages, *Infect. Immun.,* 51, 141, 1986.

73. **Brown, K. M. and Kreier, J. P.,** *Plasmodium berghei* malaria: blockage by immune complexes of macrophage receptors for opsonized plasmodia, *Infect. Immun.,* 37, 1227, 1982.

74. **Shear, H. L.,** Murine malaria: immune complexes inhibit Fc receptor-mediated phagocytosis, *Infect. Immun.,* 44, 130, 1984.

75. **Shear, H. L, Brown, J., Gysin, J., Jensen, J. B., and Tharvanij, S.,** Human and primate malarial sera inhibit Fc receptor-mediated phagocytosis, *Am. J. Trop. Med. Hyg.,* 36, 234, 1987.

76. **Ward, K. N., Warrell, M. J., Rhodes, J., Looareesuwan, S., and White, N. J.,** Altered expression of human monocyte Fc receptors in *Plasmodium falciparum* malaria, *Infect. Immun.,* 44(3), 623, 1984.

77. **Quinn, T. C. and Wyler, D. J.,** Resolution of acute malaria (*Plasmodium berghei* in the rat): reversibility and spleen dependence, *Am. J. Trop. Med. Hyg.,* 29, 1, 1980.

78. **Loose, L. D.,** Characterization of macrophage dysfunction in rodent malaria, *J. Leuk. Biol.,* 36, 703, 1984.

79. **Weindanz, W. P.,** Malaria and alterations in immune reactivity, *Br. Med. Bull.,* 38, 167, 1982.

80. **Murphy, J. R.,** Host defenses in murine malaria: analysis of Plasmodial infection-caused defects in macrophage microbicidal capacities, *Infect. Immun.,* 31(1), 396, 1981.

81. **Coleman, R. M., Prencricca, N. H., Stout, J. P., Brisette, W. H., and Smith, D. M.,** Splenic mediated erythrocyte cytotoxicity in malaria, *Immunology,* 29, 49, 1975.

82. **Eugui, E. M. and Allison, A. C.,** Malaria infections in different strains of mice and their correlation with natural killer activity, *Bull. W.H.O.,* 57 (Suppl. 1), 231, 1979.

83. **Nnalue, N. A. and Friedman, M. J.,** Evidence for a neutrophil-mediated protective response in malaria, *Parasitol. Immunol.,* 10, 47, 1988.

84. **Criswell, B. S., Butler, W. T., Rossen, R. D., and Knight, V.,** Murine malaria: the role of humoral factors and macrophages in destruction of parasitized erythrocytes, *J. Immunol.,* 107, 212, 1971.

85. **Langhorne, J., Butcher, G. A., Mitchell, G. H., and Cohen, S.,** Preliminary investigations on the role of the spleen in immunity to *Plasmodium knowlesi* malaria, *Trop. Dis. Res. Ser.,* 1, Schwabe & Co. AG, Basel, 205, 1979.

86. **Taverne, J., Dockrell, H. M., and Playfair, J. H. L.,** Killing of the malarial parasite *Plasmodium yoelii in vitro* by cells of myeloid origin, *Parasitol. Immunol.,* 4, 77, 1982.

87. **Ockenhouse, C. F. and Shear, H. L.,** Oxidative killing of the intraerythrocytic malaria parasite *Plasmodium yoelii* by activated macrophages, *J. Immunol.,* 132, 424, 1984.

88. **Ockenhouse, C. F. and Shear, H. L.,** Malaria-induced lymphokines: stimulation of macrophages for enhanced phagocytosis, *Infect. Immun.,* 42, 733, 1983.

89. **Ockenhouse, C. F., Schulman, S., and Shear, H. L.,** Induction of crisis forms in the human malaria parasite *Plasmodium falciparum* by gamma-interferon-activated, monocyte-derived macrophages, *J. Immunol.,* 133(3), 1601, 1984.

90. **Barnwell, J. W., Ockenhouse, C. F., and Knowles, D. M., II,** Monoclonal antibody OKM5 inhibits the *in vitro* binding of *Plasmodium falciparum*-infected erythrocytes to monocytes, endothelial, and C32 melanoma cells, *J. Immunol.,* 135, 3494, 1985.

91. **Perlmann, P. and Cerottini, J. C.,** Cytotoxic lymphocytes, in *The Antigens,* Vol. 5, Sela, M. Ed., Academic Press, New York, 1979, 173.

92. **Brown, J. and Smalley, M. E.,** Specific antibody-dependent cellular cytotoxicity in human malaria, *Clin. Exp. Immunol.,* 41, 423, 1980.

93. **Gilbreath, M. J., Pavanand, K., MacDermott, R. P., Phisphumvithi, P., Permpanich, B., and Wimonwattrawatee, T.,** Different spontaneous cell-mediated cytoxicity and lectin-induced cellular cytotoxicity by peripheral blood mononuclear cells from Thai adults naturally infected with malaria, *J. Clin. Microbiol.,* 17, 296, 1983.

94. **McDonald, V. and Phillips, R. S.,** Increase in non-specific antibody medicated cytoxicity in malarious mice, *Clin. Exp. Immunol.,* 34, 159, 1978.

95. **Greenwood, B. M., Oduloji, A. J., and Stratton, D.,** Lymphocyte changes in acute malaria, *Trans. R. Soc. Trop. Med. Hyg.,* 71, 408, 1975.

96. **Brown, J. and Greenwood, B. M.,** Cellular and humoral inhibition of *Plasmodium falciparum* growth *in vitro* and recovery from acute malaria, *Parasitol. Immunol.,* 7, 265, 1985.

97. **Clark, I. A., Allison, A. C., and Cox, F. E.,** Protection of mice against babesia and plasmodium with BCG, *Nature,* 259, 309, 1976.

98. **Clark, I. A. and Hunt, N. H.,** Evidence for reactive oxygen intermediates causing hemoloysis and parasite death in malaria, *Infect. Immun.,* 39, 1, 1983.

99. **Clark, I. A., Hunt, N. H., Cowden, W. B., Maxwell, L. E., and Mackie, E. J.,** Radical-mediated damage to parasites and erythrocytes in *Plasmodium vinckei* infected mice after injection of t-butyl hydroperoxide, *Clin. Exp. Immunol.,* 56, 524, 1984.

100. **Dockrell, H. M. and Playfair, J. H. L.,** Killing of blood-stage murine malaria parasites by hydrogen peroxide, *Infect. Immun.,* 39, 456, 1983.

101. **Wozencraft, A. O., Dockrell, H. M., Taverne, J., Targett, G., A. T., and Playfair, J. H. L.,** Killing of human malaria parasites by macrophage secretory products, *Infect. Immun.,* 43, 664, 1984.

102. **Makimura, S., Brinkmann, V., Mossmann, H., and Fischer, H.,** Chemiluminescence response of peritoneal macrophages to parasitized erythrocytes and lysed erythrocytes from *Plasmodium berghei*-infected mice, *Infect. Immun.,* 37, 800, 1982.

103. **Brinkmann, V., Kaufmann, S. H. E., Simon, M. M., and Fischer, H.,** Role of macrophages in malaria: O_2 metabolite production and phagocytosis by splenic macrophages during lethal *Plasmodium berghei* and self-limiting *Plasmodium yoelii* infection in mice, *Infect. Immun.,* 44, 743, 1984.

104. **Wozencraft, A. O., Croft, S. L., and Sayers, G.,** Oxygen radical release by adherent cell populations during the initial stages of a lethal rodent malarial infection, *Immunology,* 56, 523, 1985.

105. **Kharazmi, A., Jepsen, S., and Andersen, B. J.,** Generation of reactive oxygen radicals by human phagocytic cells activated by *Plasmodium falciparum, Scand. J. Immunol.,* 25, 335, 1987.
106. **Clark, I. A., Hunt, N. H., and Cowden, W. B.,** Oxygen-derived free radicals in the pathogenesis of parasitic disease, *Advances in Parasitology,* Vol. 25, 1, 1986.
107. **Friedman, M. J.,** Oxidant damage mediates variant red cell resistance to malaria, *Nature,* 280, 245, 1979.
108. **Roth, E. F., Raventos-Suarez, C., Rinaldi, A., and Nagel, R. L.,** Glucose-6-phosphate dehydrogenase deficiency inhibits *in vitro* growth of *Plasmodium falciparum, Proc. Natl. Acad. Sci. U.S.A.,* 80, 298, 1983.
109. **Roth, E. F., Shear, H. L., Constantini, F., Tanowitz, H. B., and Nagel, R. L.,** Malaria in β-thalassemic mice and the effects of the transgenic human β-globin gene and splenectomy, *J. Lab. Clin. Med.,* 111, 35, 1988.
110. **Root, R. K., Metcalf, J., Oshino, N., and Chance, B.,** H_2O_2 release from human granulocytes during phagocytosis. I. Documentation, quantitation, and some regulating factors, *J. Clin. Invest.,* 55, 945, 1975.
111. **Shear, H. L., Nolan, T., Srinivasan, R., and Ng, C.,** Recombinant gamma interferon protects mice against the lethal strain of *Plasmodium yoelii 17x, J. Leuk. Biol.,* 42, 420, 1987.
112. **Clark, I. A., Hunt, N. H., Butcher, G. A., and Cowden, W. B.,** Inhibition of murine malaria (*Plasmodium chabaudi*) *in vivo* by recombinant interferon-gamma or tumor necrosis factor, and its enhancement by butylated hydroxyanisole, *J. Immunol.,* 10, 3493, 1987.
113. **Clark, I. A., Virelizier, J., Carswell, E. A., and Wood, P., R.,** Possible importance of macrophage-derived mediators in acute malaria, *Infect Immun.,* 31, 1058, 1981.
114. **Taverne, J., Dockrell, H. M., and Playfair, H. L.,** Endotoxin-induced serum factor kills malarial parasites *in vitro, Infect. Immun.,* 33, 83, 1981.
115. **Rzepczyk, C. M. and Clark, I. A.,** Demonstration of a lipopolysaccharide-induced cytostatic effect on malarial parasites, *Infect. Immun.,* 33, 343, 1981.
116. **Taverne, J., Depledge, P., and Playfair, J. H. L.,** Differential sensitivity *in vivo* of lethal and nonlethal malarial parasites to endotoxin-induced serum factor, *Infect. Immun.,* 37, 927, 1982.
117. **Haidaris, C. G., Haynes, J. D., Meltzer, M. S., and Allison, A. C.,** Serum containing tumor necrosis factor is cytotoxic for the human malaria parasite *Plasmodium falciparum, Infect. Immun.,* 42, 385, 1983.
118. **Rzepczyk, C. M.,** Probable macrophage origin of the lipopolysaccharide-induced cytostatic effect on intraerythrocytic malarial parasites *(Plasmodium vinckei), Immunology,* 46, 261, 1982.
119. **Rzepczyk, C. and Clark, I. A.,** Failure of bacterial lipopolysaccharide to elicit a cytostatic effect on *Plasmodium vinckei* petteri in C3H/HeJ mice, *Infect. Immun.,* 35, 58, 1982.
120. **Cottrell, B. J., Playfair, J. H. L., and De Souza, B. J.,** Cell-mediated immunity in mice vaccinated against malaria, *Clin. Exp. Immunol.,* 34, 147, 1978.
121. **Playfair, J. H. L., De Souza, J. B., Dockrell, H. M., Agomo, P. U., and Taverne, J.,** Cell-mediated immunity in the liver of mice vaccinated against malaria, *Nature,* 282, 731, 1979.
122. **Taverne, J., Treagust, J. D., and Playfair, J. H. L.,** Macrophage cytotoxicity in lethal and non-lethal murine malaria and the effect of vaccination, *Clin. Exp. Immunol.,* 66, 44, 1986.
123. **Freeman, R. R. and Holder, A. A.,** Characteristics of the protective response of BALB/c mice immunized with a purified *Plasmodium yoelii* schizont antigen, *Clin. Exp. Immunol.,* 54, 609, 1983.
124. **Siddiqui, W. A., Taylor, D. W., Kan, S. C., Kramer, K., Richmond-Crumm, C. M., Kotani, S., Shiba, T., and Kasumoto, S.,** Immunization of experimental monkeys against *Plasmodium falciparum:* use of synthetic adjuvants, *Bull. W.H.O.,* 57 (Suppl. 1), 199, 1979.
125. **Perrin, L. H., Merkli, B., Gabra, M. S., Stocker, J. W., Chizzolini, C., and Richle, R.,** Immunization with a *Plasmodium falciparum* surface antigen induces a partial immunity in monkeys, *J. Clin. Invest.,* 75, 1718, 1985.
126. **James, S. L. and Scott, P.,** Induction of cell-mediated immunity as a strategy for vaccination against parasites, in *The Biology of Parasitism,* Alan R. Liss, New York, 1988, 249.
127. **Weiss, W. R., Sedegah, M., Beaudwin, R. L., Miller, L. H., and Good, M. F.,** CD8[+] T cells (cytotoxic/suppressors) are required for protection in mice immunized with malaria sporozoites, *Proc. Natl. Acad. Sci. U.S.A.,* 85, 573, 1988.
128. **Schofield, L., Villaquiran, J., Ferreira, A., Schellekens, H., Nussenzweig, V., and Nusenzweig, R. S.,** Gamma interferon, CD8[+] T cells and antibodies required for immunity to malaria sporozoites, *Nature,* 330, 664, 1987.
129. **Sher, A.,** Vaccination against parasites: special problems imposed by the adaptation of parasitic organisms to the host immune response, in *The Biology of Parasitism,* Alan R. Liss, New York, 1988, 169.
130. **Playfair, J. H. L. and De Souza, J. B.,** Recombinant gamma-interferon is a potent adjuvant for a malaria vaccine in mice, *Clin. Exp. Immunol.,* 67, 5, 1987.
131. **Shear, H. L.,** Variation inexpression of antibody-dependent cell-mediated cytotoxicity in rodents with malaria, *Infect. Immun.,* 56, 3007, 1988.

Chapter 5

MALARIA CRISIS FORMS: INTRAERYTHROCYTIC DEVELOPMENT DERANGEMENT

James B. Jensen

TABLE OF CONTENTS

I. IMMUNITY TO BLOOD-STAGE MALARIA INFECTIONS

A. HUMORAL IMMUNITY

It has long been recognized that *Plasmodium* parasites contain many antigens and that infections with these parasites produce abundant antibody to manifold determinants, only a few of which appear to be important to protective mechanisms.[1] Due to the intracellular nature of the infection in the intermediate host, antibody must react with the parasite in such a manner as to reduce its reproductive capacity to be of benefit in controlling the infection. In the erythrocytic cycles of asexual reproduction, only antigens exposed to the humoral elements of the immune system are susceptible to effective antibody action. This generally means that only antibody associated with the extracellular merozoites,[2,3] or neoantigens exposed on the surface of the infected erythrocyte,[4] are capable of interfering with parasite reproduction. In the case of the merozoite, associated antibody may interfere with recognition of, adherence to, and invasion into the erythrocyte — an invasion blocking, or neutralizing action, such as suggested for the ring-infected erythrocyte surface antigen (RESA), Pf-155 antigen.[5,6] This antigen is associated with the merozoite polar organelles, part of the erythrocyte-invasion machinery of the parasite,[7] and is readily demonstrated within the inner portion of the newly invaded red blood cell plasmalemma. Merozoite surface antigens have been implicated in parasite agglutination,[8] and such antibody-mediated cross-linkage demonstrably prevents completion of the invasion process.[9] There exists yet another category of potentially effective antibody to the parasite-derived or -modified antigens exposed on the surface of the infected erythrocytes (the RESA is not a true "surface" antigen since it is located below the outer layer of the plasmalemma and is only revealed by air drying the infected cells to expose the inner portions of the erythrocyte membrane[10]). Antibody to these antigens prevents or reverses sequestration of the maturing parasitized red cells to capillary endothelium, thus allowing the spleen, and possibly other tissues of the reticuloendothelial system, to remove the infected erythrocytes by phagocytosis.[11] Thus, targets and mechanisms of clinically important antibody include anti-RESA, an invasion neutralizing antibody; merozoite surface antibody which agglutinates the newly released parasites, resulting in their failure to complete invasion of the erythrocytes; and antibody to antigens on the infected erythrocytes which have two different effector mechanisms, the prevention or reversal of infected erythrocyte cytoadherence to capillary endothelium and the promotion of phagocytosis of these infected cells.

B. NONANTIBODY MECHANISMS OF PARASITE INHIBITION

These antibody-mediated processes notwithstanding, it has not been possible to describe clinical immunity to blood-stage infections strictly in terms of antimerozoite, or antiinfected erythrocyte antibody.[12] Nonantibody mechanisms, usually referred to as cell-mediated immunity, must work in concert with the humoral responses to effect an immunologic resolution of malaria infections. The antibody-dependent reversal of sequestration and subsequent opsonization-mediated phagocytosis is the most easily recognized "cellular" mechanism. Some investigators would argue that since this is antibody-dependent phagocytosis, it is not truly cell-mediated immunity.[13] In this regard, cell-mediated immunity refers to those antibody-free mechanisms which interfere with parasite reproduction. Since cytotoxic T-lymphocytes have never been implicated in malaria immunity, this leaves the retardation of intraerythrocytic parasite development often leading to moribund parasites known as crisis forms.

C. THE ORIGINAL DESCRIPTION OF CRISIS FORMS

The original description of crisis forms is found in a 1944 report by Taliaferro and Taliaferro.[14] Because few investigators are familiar with the details of this classical report and many researchers have confused any moribund intraerythrocytic malarial parasites with true crisis forms, some of the original descriptions of crisis forms, as detailed in that report, are given here. Prior to the investigation which led to the discovery of crisis forms, Taliaferro had postulated that the rate of parasite increase, and thus the degree of pathogenesis in any given malaria infection, could be predicted mathematically by knowing the average number of merozoites per cycle and the time period of schizogony. This concept came under attack by Boyd, and others, who reported that during avian malaria infections the number of merozoites per schizont was not static, but decreased as the infection progressed. Thus, in an attempt to reassess his hypothesis, Taliaferro made careful observations on parasite development during experimental infections of *Plasmodium brasilianum* in cebus and spider monkeys. In such infections, the parasitemias would usually increase steadily to some maximal value, then decrease slowly to a state of chronicity. However, in some cases the parasitemias would rise quickly, creating a state of crisis where either the animal died or began a rapid recovery. As the infection began to subside (and sometimes even if the animal eventually died) there appeared in the infected erythrocytes an intraerythrocytic derangement of parasite development. Infections with *P. brasilianum* in these monkeys normally displayed a high degree of synchrony, with schizogony being completed about noon, every 3rd day. However, during a crisis, the parasites became markedly retarded in their development, many requiring 5 d to reach segmentation, which often resulted in marked reduction in the number of merozoites per segmenter. Many parasites failed to complete their development and became distinctly deformed. The following are direct quotes from the original report:

After the parasitemia increased alarmingly it . . .

> . . . decreased markedly during some intense crises, especially in cebus and spider monkeys. At such times, some parasites, so called crisis forms, were markedly atypical and degenerate in appearance. . . . Some segmenters which were found around the time of crisis, so called crisis forms, were often difficult to classify on account of unequally divided, clumped together or otherwise atypical appearing merozoites. In fact, some of these had to be omitted because the nuclear masses were impossible to separate into recognizable nuclei. In many such forms, the cytoplasm did not stain. . . . Segmenters were not the only forms which were abnormal. However, as the cells lingered on they became increasingly abnormal. . . . Some crisis segmenters appeared to be incapable of breaking the red cell and, within the old cell, were incapable of continuing their development and, hence, degenerated. . . . In fact, all stages of the parasite, including gametocytes, could be found which looked as if they were degenerating as determined by vacuolation, abnormally stunted size, irregular contours, darkly staining pigment, which sometimes clumped prematurely, poor staining of cytoplasm, and irregular divisions of the nuclei of the schizonts.

Thus, it can be seen that crisis forms were parasites which became retarded in their rate of development, resulting in a loss of synchronous growth, some having reasonably well-defined, albeit abnormal, morphologic characteristics. These quotes were cited here because some investigators broadly define any abnormal intracellular malaria parasite as a crisis form. In fact, moribund intraerythrocytic parasites may result from many factors quite different from those described by the Taliaferros. In the years that followed the identification of crisis forms, they were regularly reported in rodent,[15] primate,[16] and even human malaria infections.[17] Although the conditions which precipitate the appearance of crisis forms are not well characterized, they are usually associated with severe infections, often accompanying rapid, nearly fatal, increases in parasitemia. The host does not always survive and the crisis forms appear just before death, or as the infection is abrogated by the host without external intervention. Many descriptions of such episodes, including those of the Taliaferros, attributed the marked decrease in parasitemia and the induction of crisis forms to immunologic mechanisms.

D. POTENTIAL MECHANISMS OF CRISIS FORM INDUCTION

Studies into potential mechanisms underlying the development of crisis forms were first introduced by Clark and colleagues who noted that rodents infected with BCG,[18] or *Coryne-bacterium parvum*,[19] were protected from normally lethal or serious infections by species of *Babesia* and *Plasmodium*. Not only did the mice survive these hemoprotozoan infections, but the parasites in the BCG-infected mice were retarded in their development, becoming moribund within their host erythrocytes. Later investigations have shown that several agents known to stimulate cellular immune mechanism or to precipitate responses to intracellular pathogens were equally efficacious in reducing the severity of experimental *Plasmodium* infections.[20,21] Since it was well known that BCG and *C. parvum* treatments trigger the release of many nonspecific factors of cell-mediated immunity, further investigations by Clark et al.[22] and Playfair's group[23] soon led to the demonstration that sera from mice infected with BCG and injected with bacterial lipopolysaccharide (LPS) contained soluble substances which could kill malaria parasites *in vitro*. Serum derived from BCG-infected animals (generally mice or rabbits), which have been further stimulated by injections with LPS induced necrosis of tumors and is potently cytotoxic to certain tumor cell lines *in vitro*,[24] is known as tumor necrosis serum TNS, and contains the monokine, tumor necrosis factor, TNF. Our basic understanding of this process is that macrophages within the BCG-infected host become "primed" by the intracellular presence of these bacteria and that upon further stimulation with LPS they undergo a series of responses leading to the release of reactive oxygen species (ROS) such as singlet oxygen, hydrogen peroxide or superoxide ions, and monokines, such as TNF. Thus, the explanation for the malaria parasite killing in BCG-infected animals was the generation of ROS to which the parasites were demonstrably sensitive.[25] This hypothesis for the induction of crisis forms *in vivo* by oxidant stress could not entirely explain the *in vitro* results where ROS are too short-lived to affect parasites incubated in serum collected from the treated animals, therefore, it was proposed that TNF was the soluble factor responsible for parasite degeneration.[26] There were several compelling lines of investigation which supported the conclusion that TNF was responsible for malaria crisis forms, specifically studies which demonstrated that antibodies to partially purified TNF could abrogate crisis form induction of BCG/LPS mouse serum[27] and experiments which suggested that the same factor in TNF-containing mouse serum was responsible for both antitumor and antiparasitic activities.[28] Thus, for a time, the induction of malaria crisis forms could be conveniently explained by the actions of ROS *in vivo*, and TNF.

E. HUMAN CRISIS FORM FACTOR (CFF)

Interest in crisis forms came into sharper focus when it was reported that a soluble factor in serum derived from individuals living in malarious regions of Sudan could induce crisis forms in cultured *Plasmodium falciparum*.[29] In these studies, *P. falciparum* cultured in the presence of the Sudanese malaria-immune serum became retarded in their development, resulting in a breakdown of their synchronous development. Following a few hours of exposure to the immune serum, development of the intraerythrocytic parasites became morphologically abnormal, appearing shrunken, pyknotic, and karyorrhexic, resulting in moribund organisms which produced no viable merozoites (Figure 1). Later studies showed that these retarded parasites were metabolically inert, failing to incorporate radiolabeled hypoxanthine into their nucleic acids (see Table 1).[30] Field studies suggested that this factor (called CFF) was associated with clinical immunity to malaria in Sudan.[31] Other investigations indicated that this factor might be part of the acquired immune response to malaria because in longitudinal studies the serum concentrations of CFF were significantly higher during the malaria transmission season in Sudan than in the same individuals during the dry season.[32]

F. ROLE OF OXIDANT STRESS IN CRISIS FORM INDUCTION

Malaria parasites are unusually sensitive to oxidant stress;[33] thus it is likely that they would be damaged by exposure to ROS if they were in the vicinity of a macrophage respiratory

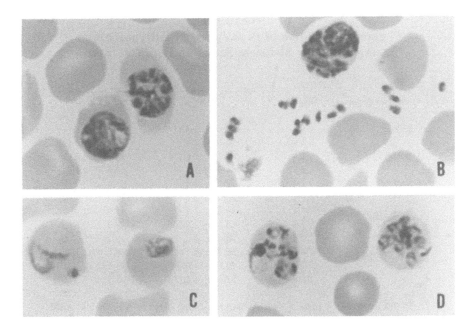

FIGURE 1. Highly synchronized cultured parasites of *Plasmodium falciparum* were grown in RPMI 1640 medium supplemented with human immune serum from Sudan or serum collected from non-immune American researchers in the expedition during serum collection. The cultures were initiated with young ring-stage parasites and grown 44 h in the presence of the test, or control sera, respectively. These photomicrographs of Giemsa-stained thin films were taken at the end of one complete cycle of intraerythrocytic development. In the nonimmune serum, the synchronous ring-stage parasite had matured to multinucleated schizonts (A), containing 16 to 24 richly stained nuclei or (B) individual merozoites, some of which are seen free of their red cell in preparation for invasion into new erythrocytes. In those cultures containing the crisis form factor serum, many of the ring-stage parasites failed to mature beyond the single nucleated trophozoite stage (C). These parasites were poorly stained, many of the parasites were pyknotic, with karyorrhexic nucleoplasm. In some parasites, there appeared abnormal merozoites which were larger in size but fewer in number when compared to parasites which developed in the nonimmune serum. In (D), the parasite to the right contains two nuclei and an undifferentiated residual mass, whereas the one on the left contains eight enlarged nuclei. These crisis form parasites (C and D) are much like those reported by Taliaferro and Taliaferro,[14] and are relatively inactive metabolically — incorporating essentially no [³H]hypoxanthine into their nucleic acids. (From Jensen, J. B. et al., *Science*, 216, 1230, 1982. With permission.)

burst. Since activated macrophages are known to be important in controlling many intracellular pathogens, principally by the generation of ROS, these observations tend to support the idea that crisis forms may result from respiratory bursts of monocytes and macrophages.[34] Infections with several species of *Plasmodium* could be affected this way, especially when infected erythrocytes circulate through lymphoid tissue containing stimulated macrophages.[35] The potential for this mechanism has been demonstrated experimentally by Clark and Hunt,[36] who induced crisis forms *in vivo* by the injection of H_2O_2-generating alloxan or H_2O_2 itself into mice infected with *P. vinckei*. This damage to the intraerythrocytic parasites was inhibited by coadministration of oxygen radical scavengers. Studies by Ockenhouse et al.[37] demonstrated that macrophages activated by interferon-γ released ROS which resulted in intraerythrocytic death of cultured *P. falciparum*. Interesting, however, was the observation in this same report that macrophages from a patient with chronic granulomatous disease (thus unable to release ROS) released a nonoxidative factor that induced crisis forms *in vitro*. Thus, it has been demonstrated that inducers of oxidant stress, such as ROS, are capable of injuring intraerythrocytic malaria parasites, and may be one mechanism involved in producing crisis forms. Despite these experimental findings, there are compelling reasons to believe that

TABLE 1
Percent Inhibition of *P. falciparum* [³H]Hypoxanthine Incorporation by Sudanese Sera with High CFF Activity

Serum donor	Dilution	% Inhibition 36 hᵃ	% Inhibition 48 hᵇ
830	1:4	70.0 ± 1.4 (2)ᶜ	89.1 ± 2.8 (2)
	1:8	54.0 ± 4.4 (2)	76.8 ± 4.0 (2)
S-81-15	1:4	88.1 ± 2.8 (2)	98.5 ± 0.3 (2)
	1:8	64.9 ± 1.8 (2)	94.6 ± 1.1 (2)
S-81-55	1:4	72.0 ± 4.4 (2)	91.4 ± 2.7 (2)
	1:8	50.1 ± 1.6 (2)	74.8 ± 1.3 (2)
S-81-132	1:4	83.5 ± 7.8 (2)	97.9 ± 0.3 (2)
	1:8	64.4 ± 2.8 (2)	92.3 ± 0.1 (2)
S-81-139	1:4	79.4 ± 4.2 (2)	96.2 ± 1.7 (2)
	1:8	56.2 ± 3.1 (2)	88.4 ± 2.1 (2)
Totalᵈ	1:4	78.6 ± 7.9 (10)	94.6 ± 4.1 (10)
	1:8	57.9 ± 6.6 (10)	85.8 ± 8.2 (10)

ᵃ Intraerythrocytic parasite development retardation, rings to schizonts, exposed 36 h to serum.
ᵇ Total parasite retardation, schizont to schizont, exposed 48 h to serum.
ᶜ Percent inhibition ± SD. Number of replicates in parentheses.
ᵈ Mean of five Sudanese sera.

oxidant stress is not the sole cause of crisis forms, especially in falciparum malaria. For example, in those cases where macrophage-generated ROSs are effective at killing pathogens, the target cells are often inside the macrophages, such as in *Toxoplasma, Leishmania, Mycobacterium* spp., etc. or are in direct contact, or close proximity to the source of the radicals. In most cases this occurs as the parasitized cells squeeze between resident histocytes within the spleen, or other reticuloendothelial tissues, or by the infiltration of mononuclear cells into tumors and other diseased tissues. However, in falciparum malaria the erythrocytes containing the mature stages are sequestered in the capillaries and do not circulate through lymphoid tissues and, although it is possible that macrophages or monocytes may adhere to the sequestered schizonts, such a juxtaposition has never been described in the thousands of microscopical examinations of vasculature occluded with parasitized erythrocytes. Thus, although crisis forms in falciparum malaria may occur by oxidant stress induced by ROS generated by monocytic cells, direct evidence in support of this hypothesis has yet to be reported. The *P. falciparum* crisis forms reported in the capillaries of patients suffering from cerebral malaria were probably induced by soluble factors which originated elsewhere since there was no evidence of leukocyte infiltration near the affected parasites.[17] Moreover, parasites exposed to ROS are usually killed quickly, thus they do not fit the description of crisis forms reported by the Taliaferros[14] or by Jensen et al.[29]

The suggestion that crisis forms in falciparum malaria may result from the generation of ROS by specific oxidases and appropriate substrates such as glucose/glucose oxidase or xanthine/xanthine oxidase combinations[25,33] led us to examine the possibility that human CFF serum might contain similar ROS-generating enzymes which reacted with constituents of the parasite culture milieu and produce sufficient oxidant stress to cause crisis forms when these sera were used in assays for CFF. Consequently, we set up the CFF assays in the presence of antioxidants, reducing agents, and scavengers of ROS such as superoxide dismutase and catalase. Since none of the 13 different reagents tested abrogated the antiparasitic action of CFF-containing sera, we concluded that it was unlikely that the *P. falciparum* crisis forms generated in culture could be attributed to oxidative mechanisms.[38] Whereas it is possible that ROS might be generated by oxidase-enzyme systems *in vivo*, and could generate crisis forms, such reactions are likely to result in significant tissue damage and thus might be of greater detriment to the host than the malaria infection. It is conceivable that such is the case and part of the immunopathology of malaria.[39]

There is one area where oxidative mechanisms, set in motion by respiratory bursts of stimulated monocytic cells, could play a central role in producing crisis forms and that is in the generation of lipid peroxidation products.[40,41] According to this model, serum lipids exposed to ROS produced by activated macrophages could undergo peroxidation and the resultant aldehydes, when incorporated into erythrocyte, or parasite lipids could disrupt parasite development, thereby, inducing the crisis forms. Such lipid peroxidation products could be stable in CFF serum and when cultured with *P. falciparum*, produce the changes consistent with crisis forms.

II. CHARACTERISTICS OF HUMAN CFF

Although malaria crisis forms have been reported in experimental infections for many years, it was the demonstration that human malaria-immune serum from Sudan that provided an opportunity to examine a "natural" serum component which induces these forms that is not produced under the somewhat harsh and "unnatural" experimental conditions; that is, crisis forms in the experimental animals have always been induced by factors not directly related to the malaria infection. It could be argued that the induction of crisis forms by BCG/LPS, or other harsh stimuli, is not related to the parasite infection or to natural immunologic mechanisms of this disease, especially since without such "artificial" stimulation crisis forms cannot be reliably precipitated. For these reasons, the discovery of CFF in human serum from malarious regions of Sudan created a link between this antiparasitic phenomenon and the infection itself. Below is a detailed description of the generation of crisis forms in *P. falciparum in vitro* by human serum from Sudan.

To effectively demonstrate crisis forms in *P. falciparum*, the cultures must by synchronized to young ring stages having a 4-h age differential. This is done by culturing the parasites according to standard methods,[42] and synchronizing them using a combination of the sorbitol[43] and the gelatin flotation methods[44] as previously described.[32] Individual sera vary greatly in the amount of CFF and thus the effect on the parasites differs from serum to serum. In sera having high titers of CFF, parasite development is retarded by greater than 90%, as determined by the incorporation of [^{3}H]hypoxanthine into parasite nucleic acids when used at 25% (1:4) concentrations in medium RPMI 1640 over a 40-h incubation period (see Table 1).[30] Giemsa-stained thin films from control cultures containing 25% nonimmune serum revealed that most of the parasites had developed to 4 to 8 nucleated stage schizonts, whereas in cultures containing immune serum the parasites were still ring stages or very small apigmented, pyknotic trophozoites that did not stain well, probably due to significant reduction in RNA in the parasite cytoplasm, Figure 1. In such experiments, as the serum concentration was reduced to 12.5 (1:8) and 6.25% (1:16) dilutions, respectively, parasite development progressed proportionally. Depending upon the amount of CFF activity in the serum, parasite morphology would vary from obviously dead organisms that were poorly stained, pyknotic, and karyorrhexic to parasites that were greatly retarded in their development, but appeared as normal ring forms or young trophozoites to parasites that were remarkably similar to the classical descriptions of crisis forms, i.e., reduced numbers of merozoites which were often larger or smaller than usual, that never produced viable merozoites. Recent studies have demonstrated that the young ring stage parasites are more susceptible to CFF than schizonts, and the action of CFF serum was reversible with parasite metabolism returning to near normal levels once the CFF-containing serum was replaced with normal serum after 8 h or less of exposure.[45] Parasite killing by ROS, in our experience,

TABLE 2
Effects of Antioxidants on Serum-Induced Developmental Retardation of
Plasmodium falciparum in vitro

Antioxidant	Concentration	Human Control ³H-Hx[a]	Human Control Nu[b]	Human CFF ³H-Hx	Human CFF Nu	Human TB ³H-Hx	Human TB Nu	Rabbit Control ³H-Hx	Rabbit Control Nu	Rabbit TNS ³H-Hx	Rabbit TNS Nu
Ascorbic acid	0.5 mM	62.4	2.2	8.7	1.1	ND	ND	53.9	2.6	6.6	1.1
α-Tocopherol	0.45 mM	70.3	2.9	11.6	1.2	ND	ND	75.5	2.9	10.8	1.5
BHT[c]	4.3 μM	97.6	4.4	12.9	1.9	ND	ND	93.4	3.9	11.2	1.5
Cystine	10.0 mM	137.5	4.3	22.1	1.8	25.4	1.8	142.5	3.7	19.6	1.6
Cysteine	10.0 mM	126.9	4.3	19.3	1.7	23.9	2.0	139.6	4.0	21.5	1.6
Glutathione	10.0 mM	111.1	4.1	16.2	1.8	18.7	1.8	109.4	3.7	11.9	1.3
Histidine	10.0 mM	98.2	4.2	14.7	1.7	ND	ND	93.3	3.9	10.4	1.4
Phenylalanine	10.0 mM	97.0	4.1	13.2	1.6	ND	ND	97.8	3.5	11.0	1.6
Tryptophan	10.0 mM	99.0	4.2	16.8	1.5	ND	ND	93.1	3.4	11.1	1.6
Tyrosine	10.0 mM	97.1	4.2	17.5	1.6	ND	ND	94.1	3.4	10.5	1.6
SOD[d]	100 μg/ml	89.9	4.0	12.9	1.8	21.2	1.6	96.0	3.6	10.1	1.3
Catalase[e]	200 μg/ml	98.2	4.1	14.3	1.9	20.8	1.8	90.9	3.8	10.1	1.4
SOD + catalase	As above	92.1	4.0	16.7	1.9	22.4	1.9	94.5	3.3	10.3	1.4
Low oxygen	1% O2	156.5	4.5	18.1	2.0	24.5	2.1	162.5	4.1	11.9	1.4
None	—	100.0	4.2	15.1	1.8	20.2	1.9	100.0	3.8	10.3	1.4

Note: ND = not determined. Na⁺ benzoate (20 mM) and mannitol (50 mM) were tested and found to be toxic to parasites at their lowest relevant antioxidant concentrations.

[a]   Percent of control ³H-Hx incorporation; n = 6 to 24; SEM were <10% of the mean.
[b]   Average number of nuclei/parasite; n = 6 to 12; SEM were <10% of the mean.
[c]   BHT = di-*tert*-butyl-ρ-hydroxytoluene.
[d]   SOD = superoxide dismutase, 2800 U/mg protein; Sigma type 1 from bovine blood.
[e]   Catalase specific activity = 33,900 U/mg protein; Sigma, from bovine liver.

is rapid and not reversible. In addition to the human serum from Sudan, we found that serum collected from class III tuberculosis patients, (those having positive PPD skin test, acid-fast bacillus stain, and chest X-ray) and sera from BCG/LPS-injected rabbits and BCG-infected mice could induce crisis forms in cultured *P. falciparum*.[46]

A. THE TUMOR NECROSIS FACTOR STORY

As has been mentioned above, the leading candidates for CFF were oxidant stress generated by ROS or products of oxidases or ROS-generated lipid peroxidation and TNF. Since our malaria parasite cultures contain no mononuclear cells to produce ROS, and since free radicals are too unstable to persist in the immune serum, the possibility that the CFF serum contains ROS-generating oxidases was investigated by adding antioxidants and free radical scavengers, such as catalase and superoxide dismutase, to CFF-containing serum (Table 2). Such treatments had no effect on reducing the CFF activity of the immune serum.[38] If crisis forms of *P. falciparum* induced *in vitro* by immune serum are not caused by the generation of ROS, it is possible that the serum contains soluble factors that, in some yet undefined way, retard parasite development — in some cases to a fatal degree. We investigated the possibility that CFF might be TNF, as suggested by others.[22,23,27,37,47] We produced TNF in rabbits by infecting them with BCG followed by injections of LPS. As expected, the serum contained TNF as determined by the standard assay for tumoricidal activity using mouse L 929 cells.[46] These rabbit sera also induced typical crisis forms in cultured *P. falciparum*. However, we noted that in examining the sera of six treated rabbits, no correlation existed in the concentrations of TNF and CFF (see Table 3). For example, one rabbit serum

TABLE 3
Plasmodium falciparum and L-M
Cell IC50s of BCG-LPS Sera

	IC50	
Rabbit serum	Parasite[a]	Tumor cell[b]
Rb-1	1:8.8	1:5408
Rb-2	1:5.9	1:2449
Rb-3	1:30.3	1:5188
Rb-4	1:5.0	1:5248
Rb-5	1:1.0	1:3266
Rb-6	1:10.7	1:6281
Total[c]	1:6.6	1:4414

[a] Concentration at which parasite incorporation of [³H]hypoxanthine is reduced by 50%.
[b] Concentration at which cell line incorporation of [³H]thymidine is reduced by 50%.
[c] Mean of six rabbits.

contained 3266 U to TNF but only 1 U of CFF (a unit being defined as the inverse of the dilution of the serum required to cause 50% inhibition of incorporation of radiolabeled precursors into the target cells), whereas a different rabbit serum contained 5188 U of TNF and 30 U of CFF. The activity against the parasites could not be predicted by the titers of TNF; thus it appears that the different cytotoxic activities are due to two different molecules. Testing of the human CFF-containing serum from Sudan in the ML 929 cell assay showed no TNF activity.[50] Further studies undertaken in collaboration with Dr. Lloyd Old of Sloan Kettering Cancer Institute (New York), demonstrated that, although mouse TNF serum contained CFF activity, a semipurified fraction containing 200-fold greater TNF activity than the serum from which it was purified contained no activity against cultured *P. falciparum*. These studies were all conducted before recombinant human TNF was available. However, with the availability of pure genetic-engineered human TNF, we were able to examine this monokine for direct activity against the parasites *in vitro*. We obtained the TNF from two different sources, Genentech and Asahi Chemical Company of Japan. These preparations of recombinant human TNF were totally inactive against the parasites even when used at concentrations up to 250,000 U/ml. Finally, monoclonal and polyclonal antibodies to human TNF were not able to reduce the activity of CFF-containing Sudanese serum.[48] Thus, despite the published reports linking crisis forms of malaria parasites to TNF, we have found that TNF-containing animal serum, generated by BCG/LPS stimulation, contains both TNF and CFF, but probably as two separate molecular entities. Moreover, pure gene-cloned human TNF is inactive against the parasites *in vitro*, and the human CFF serum cannot be inactivated using antibodies against human TNF. Since lymphotoxin, like TNF, is active against ML 929 cells and CFF is not, CFF is not lymphotoxin. These findings have been supported by similar reports,[49] as well as several personal communications from A. Cerami, I. Clark, J. Golenser, and others. Although TNF has no direct activity against *P. falciparum in vitro*, it may function in the induction of CFF activity *in vivo* since repeated injections of recombinant mouse TNF into *P. yoellii*-infected mice reduced parasitemias and prolonged survival time to this highly lethal infection.[50] The *in vivo* antiparasitic activity of TNF was also reported for preerythrocytic liver stages of *P. berghei* infections in mice.[51] In other experiments, it was found that TNF administered at a low but steady rate through the use of osmotic pumps could control experimental malaria infections in mice (L. Schofield, personal communication). These reports notwithstanding, it remains to be seen whether TNF has any direct or specific influence on natural infections of malaria.

We have also tested human interferon-α and -γ, neither of which were active against the parasites *in vitro*,[31,46] nor did interferon-γ potentiate the activity of human TNF or CFF. Interleukins I and II have also proven to be incapable of inducing *P. falciparum* crisis forms *in vitro*. Despite the fact that we have been unable to demonstrate crisis form activity in any of the characterized cytokines, we have found CFF activity in the serum of patients having active tuberculosis,[46] as well as sera from malaria immune individuals living in endemic areas of Sudan and in the serum of animals whose cell-mediated immune mechanisms had been hyperstimulated by BCG/LPS treatment. For these reasons, it is possible that CFF is a yet unidentified cytokine.

III. EPIDEMIOLOGIC CONSIDERATIONS OF CFF

We have conducting research on immunity to falciparum malaria in the central provinces of Sudan for the past 9 years. We have learned much about this disease in Sudan over these years and experience has given us a unique insight to malaria immunity in areas having different levels of endemicity. The Sudan is uniquely suited for studies on malaria for many reasons. It is the largest country in Africa with over 1 million square miles and a population estimated (since no one knows for certain) between 15 and 25 million people. Malaria is found throughout the country, but the degree of endemicity varies from extremely unstable hypoendemic to holoendemic stable. These descriptions are not sufficient to completely characterize the malaria situation because many areas change from mesoendemic unstable to hyperendemic on, more or less, regular cycles of 5 to 10 years, resulting in long cycles of transmission instability in addition to the annual dry/wet seasonal changes. For example, in 1980 to 1981, when we first began working in the Sennar region of the Blue Nile Province (about 200 mi south of Khartoum), the region was undergoing an unusually wet period after several years of moderate rains. That the degree of malaria transmission was extraordinary was not apparent to us, because of the primitiveness of the area and a paucity of weather records. We were led by the Ministry of Health officers, with whom we collaborated, to believe that the very wet conditions and high rates of malaria transmission were "normal" for this part of the country. With the clarity of years of experience in the area, we now know that our discovery of CFF occurred, in part, because we were in the right place at the right time. We did not know that the years previous to our studies had been significantly drier, that transmission rates were unusually high, that the population immunity to malaria was at a cyclic nadir, and that the coming years would be a repeat of the former drought conditions. Adding to the situation of unstable rain and malaria transmission patterns is the fact that only one paved highway exists in this vast country, ending about 225 mi south of Khartoum, making transportation and communications difficult. Moreover, the utter poverty of the country creates a situation where people living in outlying villages have essentially no primary health services and limited access to malaria chemotherapy. These environmental and societal conditions would be difficult to duplicate in other areas where malaria research is presently being conducted, which is one reason our findings with regard to CFF have not been repeated elsewhere.[52]

We originally began our studies by attempting to characterize the clinical immunity of the Sudanese villagers using standard serologic techniques and cultures of *P. falciparum* to examine antimerozoite activity in their serum. We attempted to correlate oral histories regarding malaria symptomatology with serologic and *in vitro* parasite inhibition activity. In the course of our early investigations, we found that the villages were highly malarious, with a prevalence rate in the 50 to 60% range (unusually high for this area, but we did not know that at the time). Moreover, we found that many of the villagers claimed that they never suffered from the symptoms of malaria. Upon examination of their sera, we found those who claimed to be free of malaria had low antibody titers to malaria antigens, whereas those who suffered regular bouts of malaria had significantly higher antibody concentrations. Notwithstanding these low antibody titers in the sera of the "malaria-free" donors, we found their sera to be highly inhibitory to cultured falciparum parasites, retarding intraerythrocytic parasite development, leading to crisis forms.[29] We then began to study the pattern of the presence and concentration of the factor responsible for this phenomenon in sera collected from different regions of Sudan. We found that sera collected in the hypoendemic areas, such as Khartoum and the Gezira Province, generally did not contain CFF. As we collected sera from the southern provinces, Upper Nile and Bahr El Ghazal, where malaria is hyperendemic and more malarious than Sennar, the prevalence of CFF in the sera increased significantly. Moreover, we found that those individuals who claimed to be free of malaria symptoms generally had far more CFF and lower antibody titers than individuals with histories of recurrent malaria.[31] Finally, we conducted a longitudinal study where we collected matched wet and dry season serum samples from 62 donors in the Sennar area. Of these, we obtained 19 during the next dry season (we were inhibited from collecting more samples by the arrival of Ramadan, the Moslem month of fasting). We attempted to collect a final (fourth) sample from these same individuals the next wet season, but the rains failed that year, which was the beginning of the 3-year disastrous drought in Ethiopia and Sudan between 1983 and 1986. Nonetheless, we did obtain valuable information about CFF by examining the sera we did obtain. As can be seen in Figure 2, there was a significant increase in CFF activity with the arrival of the rains and malaria transmission, although antibody titers remained somewhat stable (see Table 4).[32] What was not revealed in our report was that the sera we did collect during that final (failed) "wet" season contained no CFF activity. Moreover, we did not find CFF activity in the Sennar region for the duration of the drought, but upon the return of the rains in 1987 CFF activity was again found in the sera of many individuals living in the Sennar area. These observations have reinforced our conclusions that the presence of CFF in the serum was linked to malaria transmission.

When the rains failed in the central Sudan, we attempted to work in the most southern regions of the country where malaria is holoendemic and stable. We made two expeditions to the area of Juba, but the outbreak of civil war between the north and south of the country (which still rages today) prevented us from obtaining much data. Interestingly, however, is the fact that, of 60 samples of serum collected from adults living near Juba, only two had any CFF activity, and these had only moderate amounts. We then began to search for CFF in the serum of persons living in malarious regions outside of Sudan. We made a comparative study between Sudanese and Indonesians, in which we discovered the former to generally have low antimalarial antibody titers and high CFF, while the latter had high antibody titers and essentially no CFF.[53] We have also compared sera collected from clinically immune adults from Nigeria, Ecuador, Burkina Faso, and Liberia and found them to be free of CFF activity, like the serum samples from holoendemic Juba. Of course, these findings have required us to make some reassessments of our original claims about the role CFF plays in clinical immunity to falciparum malaria. In an attempt to determine whether CFF was a uniquely Sudanese phenomenon, we began comparative studies among different Sudanese populations living in areas of different endemicity. We also began to look at the role CFF might play in the acquisition of immunity to falciparum malaria in very young Sudanese children.

A. STUDIES OF DIFFERENT MALARIA ENDEMICITIES IN SUDAN

We conducted a comparative study between villagers living near Sennar in an area where sugar cane was grown and refined with villagers living in Sundus, a normally hyperendemic region along the Ethiopian border. Because of the economic importance of the sugar production area, and because the sugar cane fields are routinely irrigated, these villages had been under malaria control for several years. In 1984, there was a shortage of petroleum

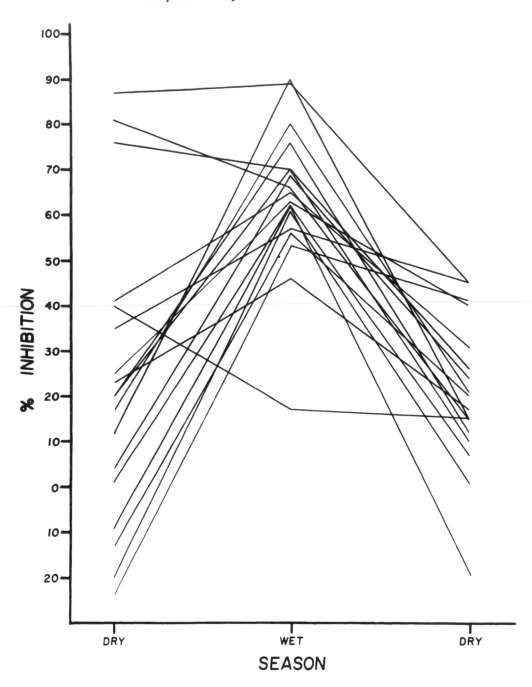

FIGURE 2.　Inhibition of intraerythrocytic development by serum samples longitudinally collected from 19 individuals living in villages near Sennar, Sudan, during the dry season, 1982; wet season, 1982; and dry season, 1983. All data points from each serum represent mean inhibition values from three experiments (SD ± 10% inhibition). (From Vande Waa, et al., *Infect. Immun.*, 45, 505, 1984. With permission.)

product in the country; thus, no insecticide for domicile spraying was delivered to the sugar production villages. The next year, the insecticide arrived and was employed in an attempt at malaria control during the irrigation season. However, in the year that the insecticide had been stored awaiting transport to the sugar refinery area, it had become inactivated (a problem with the unstable organophosphate-type insecticides); thus, it was ineffectual at preventing an epidemic of malaria. We visited these villages during the epidemic, provided treatment,

TABLE 4

Immunoglobulin Concentrations and Antimalarial Activities in Serum Samples Collected from the Same Individuals during DS and WS, 1982

Experiment	DS			WS			Significance[a]
	n	Mean	Range	n	Mean	Range	
IgG IFA[b]	62	688	20—2560	62	1152	20—5.120	NS
IgM IFA[b]	17	56	0—2560	17	72	0—320	NS
IfA IFA[b]	17	7	0—80	17	15	0—80	NS
IgG (mg/dl)[c]	62	1360	941—2275	62	1411	988—2357	NS
IgM (mg/dl)[c]	62	124	34—2275	62	129	34—2357	NS
IgA (mg/dl)[c]	62	181	69—318	62	183	83—363	NS
Merozoite invasion inhibition (5)	62	15.7	−9—51.0	62	15.4	−12.5—51.1	NS
Intraerythrocytic parasite development inhibition (%)	62	39.0	−21.1—97.0	62	66.5	16.7—93.6	(P <0.001)

[a] Student t test used to determine levels of significance; NS, not significant (P > 0.05).

[b] Reciprocal endpoint titers.

[c] Serum immunoglobulin concentrations as determined by radial immunodiffusion.

and collected data and serum samples. We were working 2 weeks later in Sundus, a village some 200 mi west of the sugar refinery, in an area which normally would have been quite malarious, but at that time was in the 2nd year of the recent drought. A comparison of the sera from these two villages for antibody and CFF activity was illuminating. The sugar refinery villages were undergoing a serious clinical epidemic of malaria with a prevalence rate of infection in schoolchildren at about 41%, with many hundreds of children and adults suffering from serious falciparum malaria. This population was poorly immune to malaria because of previous malaria control measures that had just recently broken down. In Sundus, on the other hand, no malaria control had ever been instituted, but due to the drought there was no transmission and no cases of clinical malaria, despite a prevalence of 39% parasite infections in the children. The major differences between these two villages was that residents of the one were nonimmune and exposed to high rates of malaria transmission, while in the other they were generally immune, but due to the drought were not exposed to transmission (our team slept outdoors and were not bothered by mosquitoes during the 2 weeks they were in the village). Sera from the sugar refinery villages was low in antimalarial antibody, but high in CFF, whereas the sera from Sundus had reasonably high titers of malarial antibody, but essentially no CFF (Vande Waa and Jensen, manuscript in preparation). Moreover, there was a significant negative correlation between antimalarial antibody titers and serum CFF concentration. These data strongly suggest that CFF is found in poorly immune individuals during periods of high malaria transmission pressures, but is not found in individuals reasonably immune to malaria. These observations would support our previous experience where we were unable to demonstrate CFF in sera from Indonesia, Nigeria, and other malarious areas. Moreover, they support the findings of Marsh et al.,[52] who reported that they were unable to demonstrate CFF in the sera of immune individuals from Gambia.

B. CFF AND THE ACQUISITION OF MALARIAL IMMUNITY

We conducted a 2-year study on changes in serum immune factors and the acquisition of malaria in young children in the area of Damazin, southern Blue Nile Province, Sudan. This area is hyperendemic for falciparum malaria, and significantly more malarious than the Sennar area of our initial studies. We collected 400 serum samples from children between the ages of 6 and 60 months with some 100 samples from the parents of these children. In addition to these familial samples, we collected another 200 from primary schoolchildren between ages 7 and 10 years. We also collected 65 matched neonate (umbilical cord) and mother's sera. Although we are still analyzing these sera, we have completed the CFF assays and have determined the malaria antibody titers by IFA and RESA,[5] the latter being considered a reasonable index of antimerozoite antibody activity. When we began this study, we predicted on the basis of our previous studies comparing Sudanese to Indonesian immunity to malaria where the former appeared to be more clinically immune than the latter; the one having high titers of CFF which was highly lethal to the parasites compared to the high antibody concentrations in the Indonesians who suffered from malaria splenomegaly and repeated infections, that the young children would have high titers of ineffectual malarial antibody, which would gradually be replaced with CFF as immunity to malaria became more evident with age. Our first observations demonstrated that although some new mothers had reasonable CFF titers (many had none), CFF did not pass through the placenta, since only a few of the neonates had CFF activity in their sera and there was no relationship between CFF activity of the mother's, vs. the neonate's sera. Such was not the case regarding antimalarial IgG titers which showed a high degree of correlation. (Vande Waa, and Jensen, manuscript in preparation). Unlike the sera we obtained in Sennar area, the adult sera from Damazin had essentially no CFF activity (much like the samples we collected in Juba), but they showed a significant age-related increase in RESA antibody titers. The children, on the other hand had an interesting pattern of CFF activity, 55 of 110 children under the age of 13 years had CFF activity of greater than 25%. These data are more striking when comparison is made by ascending age. For example, progressing from age 1 to 13 years, the average CFF activity in the pediatric sera was 62, 42, 47, 35, 28, 35, 30, 12, 12, 12, 8, 14, and 5% respectively. The correlation coefficient revealed these data to be highly significant (p <0.001). These figures are even more impressive when plotted against the serum titers to RESA-IFA antibodies, which show progressive increases in these same children. The RESA-IFA titers from age 1 through 13 are 0, 0, 6, 6, 6, 16, 8, 48, 24, 64, 24, and 48 (p <0.05). Thus, it can be seen that as immunity to falciparum malaria increases in this population, RESA-IFA titers also increase, but CFF activity, which is high in the nonimmune children, decreases. These data are reinforced by our findings that of 103 children, between 6 and 18 months, who were born during the drought years and had yet to be exposed to much malaria were without appreciable CFF or RESA-IFA antibodies (Jensen et al., research in progress). In summary, our findings have demonstrated that there exists a negative correlation between titers of malarial antibody and age-related acquisition of immunity to falciparum malaria and serum concentrations of CFF in Sudan. Moreover, no CFF activity is found in sera from individuals not exposed to malaria, but is readily demonstrable in sera of nonimmune or semiimmune individuals who are exposed to high malaria transmission. Our former conclusion that CFF was strongly associated with clinical immunity to malaria in Sudan was based upon a faulty assumption that the population we originally investigated was clinically immune to the disease, and this conclusion must now be modified in view of the present data.

IV. PURIFICATION OF CFF

Attempts to purify CFF from human malaria immune serum are presently under way. We have not yet succeeded in obtaining sufficient human serum to characterize this factor, but data obtained so far have been encouraging. We have discovered that CFF serum contains two distinct fractions with anti-falciparum activity *in vitro*. One fraction is lipid rich and is obtained by precipitation of serum lipids by dextran sulfate treatment or by dialysing the serum against distilled water. The lipids and lipoproteins precipitate out in both of these treatments and these lipid-rich fractions from human CFF serum, and rabbit BCG/LPS treated animals are inhibitory to parasite development. After treatment to remove the lipids, lipid-depleted portions of pooled immune and nonimmune human sera were fractionated on S-300. Two fractions from the immune serum pool, 45 to 60 kDa and <300 kDa, were inhibitory to parasite development. The larger molecule may be a polymer of the smaller unit. The S-300 fraction containing crisis form activity could not be separated from serum albumin using an anionic exchange FPLC column, but attempts at further purification using cationic exchange columns have so far been encouraging. The material thus far obtained still contains sufficient albumin to obscure any distinct bands of PAGE-gels (since the molecular weight of the active material is essentially the same as albumin), but the specific activity of the semipurified factor is greater than the serum from which it was derived. It is possible that the lipid-rich fraction represents peroxidation of serum lipids by ROS created *in vivo* by macrophage respiratory bursts, although it is interesting to note that the total serum lipid content from the BCG/LPS treated animals is many times greater than found in sera obtained before treatment, suggesting that the activation of the macrophages releases TNF which increases the lipids[54] in the plasma, which subsequently undergo peroxidation by the concomitant release of ROS.[41] The lipid-depleted portion of the CFF serum may contain a specific parasite inhibiting cytokine or it may contain peroxidized fatty acids in close association with the serum albumin.

ACKNOWLEDGMENTS

The research conducted by the author and his colleagues was sponsored by grants from The National Institutes of Health, NIAID, AI 16312, AI 22905-03, and the Thrasher Research Fund. Collaboration with numerous Sudanese scientists, especially Drs. Mustafa, A. S. Akood, and R. Bayoumi is acknowledged.

REFERENCES

1. **Heidrich, H.-G.**, *Plasmodium falciparum* antigens as target molecules for a protective immunization against malaria: an up-to-date review, *Zeitschrift Parasitenkunde*, 72, 1, 1986.
2. **Cohen, S. and Butcher, G. A.**, Properties of protective malarial antibodies, *Immunology*, 19, 369, 1970.
3. **Epstein, N., Miller, L. H., Kaushel, D. C., Udeinya, I. J., Rener, J., Howard, R. J., Asofsky, R., Aikawa, M. A., and Hess, R. L.**, Monoclonal antibodies against a specific surface determinant of malarial (*Plasmodium knowlesi*) merozoites block erythrocyte invasion, *J. Immunol.*, 127, 212, 1981.
4. **Udeinya, I. J., Miller, L. H., McGregor, I. A., and Jensen, J. B.**, *Plasmodium falciparum* strain specific antibody blocks binding of infected erythrocytes to amelanotic melanoma cells, *Nature*, 303, 429, 1983.
5. **Wahlgren, M., Bjorkman, A., Perlmann, H., Berzins, K., and Perlmann, P.**, Anti-Plasmodium falciparum antibodies acquired by residents in a holoendemic area of Liberia during development of clinical immunity, *Am. J. Trop. Med. Hyg.*, 35, 22, 1986.
6. **Wahlin, B., Wahlgren, M., Perlmann, H., Berzins, K., Bjorkman, A., Patarroyo, M. E., and Perlmann, P.**, Human antibodies to a *Mr* 155,000 *Plasmodium falciparum* antigen efficiently inhibit merozoite invasion, *Proc. Natl. Acad. Sci. U.S.A.*, 81, 7912, 1984.
7. **Brown, G. V., Culvenor, J. G., Crewther, P. E., Bianco, A. E., Coppel, R. L., Saint, R. B., Stahl, H. D., Kemp, D. J., and Anders, R. F.**, Localization of the ring-infected erythrocyte surface antigen (RESA) of *Plasmodium falciparum* in merozoites and ring-infected erythrocytes, *J. Exp. Med.*, 162, 774, 1985.
8. **Green, T. J., Morhardt, M., Brackett, R. G., and Jacobs, R. L.**, Serum inhibition of merozoite dispersal from *Plasmodium falciparum* schizonts: indicators of immune status, *Infect. Immun.*, 31, 1203, 1981.
9. **Miller, L. H., Aikawa, M. A., and Dvorak, J. A.**, Malaria (*Plasmodium knowlesi*) merozoites: immunity and the surface coat, *J. Immunol.*, 114, 1237, 1975.
10. **Perlmann, H., Berzins, K., Wahlgren, M., Carlsson, J., Bjorkman, A., Patarroyo, M. E., and Perlmann, P.**, Antibodies in malarial sera to parasite antigens in the membrane of erythrocytes infected with early stages of *Plasmodium falciparum*, *J. Exp. Med.*, 159, 1686, 1984.
11. **Udeinya, I. J., Schmidt, J. A., Aikawa, M. A., Miller, L. H., and Green, I.**, *Falciparum*-infected erythrocytes specifically bind to cultured human endothelial cells, *Science*, 213, 1981.
12. **Cohen, S. and Lambert, P. H.**, Malaria, in *Immunology of Parasitic Infections*, Cohen, S. and Warren K. S., Eds., Blackwell Scientific, Oxford, 1982, chap. 14.
13. **Clark, I. A.**, Cell-mediated immunity in protection and pathology of malaria, *Parasitol. Today*, 3, 300, 1987.
14. **Taliaferro, W. H. and Taliaferro, L. G.**, The effect of immunity on the asexual reproduction of *Plasmodium brasilianum*, *J. Infect. Dis.*, 75, 1, 1944.
15. **Barnwell, J. W. and Desowitz, R. S.**, Studies on parasitic crisis in malaria. I. Signs of impending crisis in *Plasmodium berghei* infections of the white rat, *Ann. Trop. Med. Parasitol.*, 71, 429, 977.
16. **David, P. H., Hommel, M., Miller, L. H., Udeinya, I. J., and Oligino, L. D.**, Parasite sequestration in *Plasmodium falciparum* malaria: spleen and antibody modulation of cytoadherence of infected erythrocytes, *Proc. Natl. Acad. Sci., U.S.A.*, 80, 5075, 1983.
17. **MacPherson, G. G., Warrell, M. J., White, N. J., Looareesuwan, S., and Warrell, D. A.**, Human cerebral malaria. A quantitative ultrastructural analysis of parasitized erythrocyte sequestration, *Am. J. Pathol.*, 119, 385, 1985.
18. **Clark, I. A., Allison, A. C., and Cox, F. E. G.**, Protection of mice against *Babesia* ssp. and *Plasmodium* ssp. with BCG, *Nature*, 259, 509, 1976.
19. **Clark, I. A., Cox, F. E. G., and Allison, A. C.**, Protection of mice against *Babesia* spp. and *Plasmodium* spp. with killed *Corynebacterium parvum*, *Parasitology*, 74, 9, 1977.
20. **Clark, I. A.**, Resistance to *Babesia* spp. and *Plasmodium* sp. in mice treated with an extract of *Coxilla burneti*, *Infect. Immun.*, 24, 319, 1979.

21. **Clark, I. A.,** Protection of mice against *Babesia microti* with cord factor, COAM, zymosan, glucan, *Salmonella*, and *Listeria*, *Parasitol. Immunol.*, 1, 179, 1979.

22. **Clark, I. A., Virelizier, J.-L., Carswell, E. A., and Wood, R. P.,** Possible importance of macrophage-derived mediators in acute malaria, *Infect. Immun.*, 32, 1058, 1981.

23. **Taverne, J., Dockrell, H. M., and Playfair, J. H. L.,** Endotoxin-induced serum factor kills malarial parasites, *in vitro*, *Infect. Immun.*, 33, 83, 1981.

24. **Carswell, E. A., Old, L. J., Kassel, R. L., Green, S., Fiore, N., and Williamson, B.,** An endotoxin-induced serum factor that causes necrosis of tumours, *Proc. Natl. Acad. Sci. U.S.A.*, 72, 3666, 1975.

25. **Rzepczyk, C. M., Saul, A. J., and Ferrante, A.,** Polyamine oxidase-mediated intraerythrocytic killing of *Plasmodium falciparum:* evidence against the role of reactive oxygen metabolites, *Infect. Immun.*, 43, 238, 1984.

26. **Clark, I. A.,** Does endotoxin cause both the disease and parasite death in acute malaria and babesiosis?, *Lancet*, ii, 75, 1978.

27. **Haidaris, C. G., Haynes, J. D., Meltzer, M. S., and Allison, A. C.,** Serum containing tumor necrosis factor is cytotoxic for the human malaria parasite *Plasmodium falciparum*, *Infect. Immun.*, 42, 385, 1983.

28. **Wozencraft, A. O., Dockrell, H. M.,. Taverne, J., Targett, G. A. T., and Playfair, J. H. L.,** Killing of human malaria parasites by macrophage secretory products, *Infect. Immun.*, 43, 664, 1984.

29. **Jensen, J. B., Boland, M. T., and Akood, M. A.,** Induction of crisis forms in cultured *Plasmodium falciparum* with human immune serum from Sudan, *Science*, 216, 1230, 1982.

30. **Jensen, J. B., Boland, M. T., Hayes, M., and Akood, M. A. S.,** *Plasmodium falciparum:* rapid assay for *in vitro* inhibition due to human serum from residents of malarious areas, *Exp. Parasitol.*, 54, 416, 1982.

31. **Jensen, J. B., Boland, M. T., Allan, J. S., Carlin, J. M., Vande Waa, J. A., Divo, A. A., and Akood, M. A. S.,** Association between human serum-induced crisis forms in cultured *Plasmodium falciparum* and clinical immunity to malaria in Sudan, *Infect. Immun.*, 41, 1302, 1983.

32. **Vande Waa, J. A., Jensen, J. B., Akood, M. A. S., and Bayoumi, R.,** Longitudinal study on the *in vivo* immune response to *Plasmodium falciparum* in Sudan, *Infect. Immun.*, 45, 505, 1984.

33. **Dockrell, H. M. and Playfair, J. H. L.,** Killing of *Plasmodium yoelii* by enzyme-induced products of the oxidative burst, *Infect. Immun.*, 43, 451, 1984.

34. **Allison, A. C. and Eugui, E. M.,** The role of cell-mediated immune responses in resistance to malaria, with special reference to oxidant stress, *Annu. Rev. Immunol.*, 1, 361, 1983.

35. **Playfair, J. H. L., Dockrell, H. M., and Taverne, J.,** Macrophages as effector cells in immunity to malaria, *Immunol. Lett.*, 11, 233, 1985.

36. **Clark, I. A. and Hunt, N. H.,** Evidence for reactive oxygen intermediates causing hemolysis and parasite death in malaria, *Infect. Immun.*, 39, 1, 1983.

37. **Ockenhouse, C. F., Schulman, S., and Shear, H. L.,** Induction of crisis forms in the human malaria parasite *Plasmodium falciparum* by interferon-activated, monocyte-derived macrophages, *J. Immunol.*, 133, 1601, 1984.

38. **Geary, G. T., Boland, M. T., and Jensen, J. B.,** Antioxidants do not prevent the *in vitro* induction of *Plasmodium falciparum* crisis forms by human malaria immune, TB of rabbit TNF sera, *Am. J. Trop. Med. Hyg.*, 35, 704, 1986.

39. **Stocker, R., Hunt, N. H., Buffington, G. D., Weidman, M. J., Lewis-Hughes, P. H., and Clark, I. A.,** Oxidative stress and protective mechanisms in erythrocytes in relation to *Plasmodium vinckei* load, *Proc. Natl. Acad. Sci. U.S.A.*, 82, 548, 1985.

40. **Clark, I. A., Hunt, N. H., Cowden, W. B., Maxwell, L. E., and Mackie, E. J.,** Radical-mediated damage to parasites an erythrocytes in *Plasmodium vinckei* infected mice after injection *t*-butyl hydroperoxide, *Clin. Exp. Immunol.*, 56, 524, 1984.

41. **Clark, I. A., Butcher, G. A., Buffington, G. D., Hunt, N. H., and Cowden, W. B.,** Toxicity of certain products of lipid peroxidation to the human malaria parasite *Plasmodium falciparum*, *Biochem. Pharmacol.*, 36, 543, 1987.

42. **Jensen, J. B. and Trager, W.,** *Plasmodium falciparum* in culture: use of out dated erythrocytes, and description of the candle jar method, *J. Parasitol.*, 63, 883, 1977.

43. **Lambros, C. and Vanderberg, J. P.,** Synchronization of *Plasmodium falciparum* erythrocytic stages in culture, *J. Parasitol.*, 65, 418, 1979.

44. **Jensen, J. B.,** Concentration from continuous culture of erythrocytes infected with trophozoites and schizonts of *Plasmodium falciparum*, *Am. J. Trop. Med. Hyg.*, 27, 1274, 1978.

45. **Carlin, J. M. and Jensen, J. B.,** Stage- and time-dependent effects of crisis form factor on *Plasmodium falciparum in vitro*, *J. Parasitol.*, 72, 852, 1986.

46. **Carlin, J. M., Jensen, J. B., and Geary, T. G.,** Comparison of inducers of crisis forms in *Plasmodium falciparum in vitro*, *Am. J. Trop. Med. Hyg.*, 34, 668, 1985.

47. **Playfair, J. H. L., Taverne, J., and Matthews, N.,** What is tumor necrosis factor really for?, *Immunol. Today*, 5, 165, 1984.

48. **Jensen, J. B., Vande Waa, J. A., and Karadsheh, A. J.,** Tumor necrosis factor does not induce *Plasmodium falciparum* crisis forms, *Infect. Immun.,* 55, 1722, 1987.
49. **Hviid, L., Reimer, C. M., Theander, T. H., Jepsen, S., and Bendtzen, K.,** Recombinant human tumor necrosis factor is not inhibitory to *Plasmodium falciparum in vitro, Trans. R. Soc. Trop. Med. Hyg.,* 82, 48, 1988.
50. **Taverne, J., Tavernier, J., Fiers, W., and Playfair, J. H. L.,** Recombinant tumor necrosis factor inhibits malaria parasites *in vivo* but not *in vitro, Clin. Exp. Immunol.,* 67, 1, 1987.
51. **Schofield, L., Ferreira, A., Nussenzwig, V., and Nussenzweig, R. S.,** Antimalarial activity of alpha tumor necrosis factor and gamma-interferon, *Fed. Proc., Fed. Am. Soc. Exp. Biol.,* 46, 760, 1987.
52. **Marsh, K., Otoo, L., and Greenwood, B. M.,** Absence of crisis form factor in subjects immune to *Plasmodium falciparum* in The Gambia, West Africa, *Trans. R. Soc. Trop. Med. Hyg.,* 81, 514, 1987.
53. **Jensen, J. B., Hoffman, S. L., Boland, M. T., Akood, M. A. S., Laughlin, L. W., Kurniawan, L., and Marwoto, H. A.,** Comparison of immunity to malaria in Sudan and Indonesia: crisis-form versus merozoite-invasion inhibition, *Proc. Natl. Acad. Sci. U.S.A.,* 81, 922, 1984.
54. **Beutler, B. and Cerami, A.,** Cachectin and tumor necrosis factor as two sides of the same biological coin, *Nature,* 320, 584, 1986.

Chapter 6

RELATIONSHIPS BETWEEN INFLAMMATION AND IMMUNOPATHOLOGY OF MALARIA

I. A. Clark and G. Chaudhri

TABLE OF CONTENTS

I. INTRODUCTION

As the dominant infectious disease of tropical countries, human malaria has an immediacy that earns it much attention in the clinical literature. It is also an unlikely disease, and as such is scientifically fascinating. Protozoan parasites, in very small numbers, somehow manage to injure a range of tissues and upset cellular biochemistry in others at a time when their invasiveness is limited to red cells.

The recognition and cloning of polypeptide lymphokines and monokines have transformed thinking in many fields, including the mediation of inflammation and the host response to infectious agents. In this chapter, we reason that the pattern of malarial disease within man, and across the other species, can be understood only as a host-mediated disease, with a central role being played by these polypeptide mediators released from leukocytes that have been influenced by material released from malarial parasites. These same mediators, at lower concentrations, are central to the induction of the cell-mediated immune response as well as being released during inflammation. Their secretion in toxic amounts is what, we argue, causes much of the illness and pathology seen in malaria-infected individuals. Thus, malarial pathology is true immunopathology. In helping to unmask this concept, malaria has served as a useful model for understanding the pathogenesis of the acute viral and bacterial diseases with which its clinical diagnosis is easily confused.

II. MALARIA AS A DISEASE

A. HUMAN MALARIA INFECTION

When someone never before exposed to malaria becomes infected, he remains apparently healthy until blood stages of the parasite develop and begin to build up in numbers. While these parasites are still very scarce (typically 50 to 100 parasites per cubic millimeter of blood, or 0.001 to 0.002% of red cells invaded[1]) a general malaise and fever, that is in no way distinctive for malaria, develops.

The illness typically progresses to headaches, myalgias, nausea with vomiting, fever, and rigors, but to nothing that can be diagnosed with certainty as caused by malarial parasites; therefore, a viral infection, such as influenza, is often assumed.[2] The clinical picture can be very diverse and misleading.[3] In falciparum malaria, with its more rapid parasite multiplication and adherence of red cells containing mature stages of the parasite to vascular endothelium, the illness can progress to any of a series of life-threatening manifestations. As reviewed recently,[4-6] these changes include cerebral malaria, hypoglycemia, pulmonary edema, circulatory collapse, severe anemia, thrombocytopenia, acute renal tubular necrosis, and fetal death.

The mere presence of blood stages of the parasite does not cause illness — they must first burst from their red cells. Over 100 years ago, Golgi[7] recorded that onset of paroxysms in synchronized infections correlates with the developmental stages present in erythrocytes, and information on the precise timing of these events has been available for nearly 50 years.[8] It is not surprising that analogies from bacteriology, such as clostridial toxicity, led to the concept of a malarial toxin being released, along with the new generation of merozoites, from bursting red cells.[9,10] No directly toxic malarial product has, however, been demonstrated.

As noted, the onset of clinical malaria is a remarkably nonspecific event, so much so that its precise recognition can only be a laboratory diagnosis based on detection of parasites on blood smears. How is this to be interpreted in endemic areas where thick smears from many individuals are positive regardless of clinical context? This complexity arises because a series of untreated bouts of malaria bring about a state of tolerance to parasitemias that would be a cause for concern in a previously unexposed individual. Evidently this can develop even during a single infection, since Kitchen[1] reported a series of 100 cases of vivax malaria in which the mean parasitemia at spontaneous termination of symptoms was over five times as great as that required to generate the original onset of illness. Repeated exposure enhances the phenomenon, with children from endemic regions being evidently healthy even though they are harboring 1000 times the parasite load that would confirm a diagnosis of malaria in a person newly arrived from a malaria-free area.[11,12]

While recent clinical investigations[5,6] have been thorough, there is a physical limit to what can be done to investigate the human disease. Parallels and differences between these infections and those that occur in animals have been very instructive in approaching an understanding of the pathophysiology of the human malarias. The relevant aspects of these infections are summarized in the next section.

B. MALARIA IN NON-HUMAN HOSTS
1. Monkey Malaria

The most striking contrast in monkeys and human beings experiencing malaria for the first time is that monkeys remain alert, and eat and drink, until relatively large numbers of erythrocytes are parasitized. As noted earlier, people are very different since they can become quite ill when parasites are rare enough to be difficult to find even on thick smears. This contrast is not because monkey parasites are innocuous — it also occurs when *Aotus* sp. are infected with *Plasmodium falciparum* of human origin.[13,15] Once these monkeys do become ill, however, they can suffer the same erythrophagocytosis of unparasitized red cells,[13] thrombocytopenia,[14] hypoglycemia,[14] and pulmonary edema[15] as their human counterparts. Furthermore, knob-bearing red cells adhere to vessel walls as effectively in *Aotus* sp. as they do in human patients, but in monkeys this adherence is not associated with any particular pathology.[13,16] In particular, there is no predilection for the cerebral vasculature,[17,13] which presumably accounts for the absence of cerebral malaria in these animals.

Obvious similarities with malaria-tolerant children are (1) the innate resistance of *Aotus* sp. to whatever it is in *P. falciparum* that causes illness in previously unexposed human subjects and (2) the harmless sequestration pattern of knob-bearing erythrocytes. We have previously suggested[18] that the link between these two groups may prove to be their capacity to withstand parenteral administration of bacterial endotoxin (synonym: lipopolysaccharide, or LPS); this, like malarial tolerance, is innate in monkeys (reviewed in Reference 18) and acquired in people who have recently experienced malaria.[19,20] These arguments are expanded in Section IV.A in terms of the monokines that mediate endotoxicity also being released into the circulation in malaria.

2. Malaria and Babesiosis in Other Animals

In mice, as in monkeys, it is usual to see high parasite densities in animals infected with malaria for the first time, illness not being observed (with the exception of *Plasmodium berghei* ANKA strain, as discussed in Section III.B) until parasitemias exceed 60%.[18] The same is true of *Babesia* sp., including *B. microti*, the parasite that causes onset of Nantucket Fever, a human disease, at very low parasite loads.[21] The picture is the same as that seen when people and Aotus monkeys infected with *P. falciparum* are compared. What emerges is that mice are innately very tolerant of malaria and babesia, not that the species that parasitize mice are less toxic.

As well as tolerating malaria well, mice are innately much more tolerant of both endotoxin[22] and the monokine, tumor necrosis factor (TNF),[23] than are human subjects.[24,25] Nevertheless, once illness supervenes the malarial mouse provides a useful model for a surprisingly high proportion of the pathological changes seen in human malaria, including, in *P. vinckei*-infected animals, neutrophil accumulations in pulmonary vasculature, increased plasma lactate levels, hypoglycemia,[26] dyserythropoiesis, erythrophagocytosis,[27] and fetal death.[28]

Rats present the same picture, being innately much more tolerant to malaria,[29] endo-toxin,[18] and TNF[30] than is man. Birds and reptiles also conform to this pattern with extremely high tolerances to endotoxin[18,31] and malaria parasites.[32,33] No information is available on reptiles, but chicken macrophages have proved refractory to inducers of TNF (B. Beutler, personal communication). At the other end of the scale are cattle, which are innately very sensitive to endotoxin,[31] TNF (K. McCullough, personal communication), and *Babesia* sp. parasites.[34] As this pattern predicts, changes such as pulmonary edema of the type seen in human *P. falciparum* infection occur at very low parasitemias in these animals.[35] These sets of observations strongly support our concept, expanded in Section III.A, that monokines, largely TNF, mediate malarial illness and pathology.

III. INFLAMMATION AND ITS MEDIATORS

We are all familiar with the classical redness, swelling, heat, and pain that occurs when we cut or burn a finger. It has been known for some time that the essential events behind these changes are the attraction of leukocytes to the area and the secretion by these cells of a range of bioactive soluble factors, conveniently termed the mediators of inflammation. We shall argue in this chapter that malaria and clinically similar diseases can best be understood as systemic inflammation and that the changes observed are largely the effects of circulating inflammatory mediators. This scheme is essentially as proposed by Maegraith[36] 40 years ago, updated through application of the recombinant tools that molecular biology now provides.

Broad accounts of the events of inflammation are given by Sarson and Hensen[37] and Moore and Weiss,[38] and Nathan[39] has recently compiled a current list of the mediators that macrophages potentially can contribute to these processes. We shall restrict our comments to those mediators that can plausibly be implicated in malaria. To provide a focus to this chapter we shall mainly discuss TNF and those mediators that amplify, duplicate, or inhibit its activities. As we do so, the role of lymphokines, and thus the involvement of T cells, will become evident. Thus, these changes are immunopathology, as surely as are the pro-cesses mediated more traditionally by antibody and complement. We note briefly here that lymphotoxin, a molecule of T cell origin, is not yet explored in the context of malaria. Lymphotoxin and TNF have about 35% homology in amino acid sequence and compete for the same cell-surface receptors. As yet, little is documented on the role of lymphotoxin in inflammation. Its properties and relationship to TNF have recently been reviewed by Gardner and co-workers.[40]

A. TUMOR NECROSIS FACTOR AND MALARIA

We originally proposed a link between TNF and malaria to help us understand why malaria and babesia parasites degenerate inside circulating red cells, having observed this phenomenon not only during illness crisis[41] (the timing of the original "crisis forms" in monkey malaria[42]) but also when mice had been pretreated, some weeks earlier, with a range of different macrophage activators.[43-45] These agents, selected for their capacity to protect against experimental tumors, also sensitized mice to the harmful effects of endotoxin, a property we found they shared with malaria and babesia infections.[46] These macrophage activators,[47,48] including malaria,[49] all proved to prime mice for endotoxin-triggered release of TNF. Thus we proposed[46,49] that, at schizogony, something functionally like endotoxin induced the release of TNF and other monokines and that these mediators led to both crisis forms and the illness and pathology of malaria.

1. TNF and Malaria Parasites

Several groups have now shown that recombinant TNF will inhibit malaria parasites *in vivo*[50,51] (with early onset of crisis forms[51]) but not *in vitro*.[50,52] Since there are many more interactive components *in vivo*, the differences in these results may be due to the ability of TNF to increase the sensitivity of leukocytes to agents that induce them to secrete reactive oxygen species,[53,54] and thus cause lipid peroxidation.[55] Peak *P. chabaudi* parasitemias are higher when mice are fed a diet rich in butylated hydroxyanisole, a phenolic compound with widespread commercial use as an antioxidant to retard lipid peroxidation.[51] There is recent evidence[56] that oxidatively stressed parasitized red cells contain concentrations of aldehydic products of lipid peroxidation that are directly toxic to malaria parasites *in vitro*.[57] A detailed discussion of the relationships between TNF, reactive oxygen species, and malaria crisis forms are the subject of another chapter of this book.

2. TNF and Malarial Pathology

a. Reproducing Malaria by Injecting TNF

On the basis of the limited information available 1 decade ago, we proposed that monokines, including TNF, mediated much of the pathology of endotoxicity and malaria.[46,49,58] The parallels have since become much more obvious, with many of the changes seen in experimental endotoxicity — acute renal tubular necrosis,[59] hypoglycemia,[60] hyperinsulinemia,[61] reduced hepatic gluconeogenesis,[60] low wedge pressure pulmonary edema with adherent neutrophils (adult respiratory distress syndrome; ARDS),[62] high serum lactate,[63] hypertriglyceridemia,[64] erythrophagocytosis,[65] and abortion[66] — being described, as reviewed in References 4 to 6, in acute falciparum malaria. Antibody to mouse TNF has since been shown to protect mice against endotoxin,[67] and the list of pathology present in both endotoxicity and malaria (reproducible by injecting TNF) is rapidly growing. In a malarial context we have recently reported that amounts of recombinant human TNF that were harmless to controls reproduced in mice with low parasitemias of *P. vinckei* the pulmonary neutrophil margination (Figure 1A), liver injury (Figure 1B), and hypoglycemia (Figure 2) seen in terminal illness with this infection.[26] Our current studies with recombinant TNF[27,28] have also reproduced erythrophagocytosis (Figure 3A), dyserythropoiesis (Figure 3B), and fetal death, all of which occur in human (reviewed in Reference 5) and mouse malaria.[27,28] Other changes that are recognized in malarial patients and have been reproduced in experimental animals by injecting human TNF include elevated blood lactate, diarrhea, hypotension, adrenal necrosis, and hyperinsulinemia.[68-70] Likewise, the clinical side effects observed in tumor patients given infusions of recombinant TNF are classical for malaria onset and include fever, rigors, headaches, myalgias, nausea with vomiting, and thrombocytopenia.[25]

b. Presence of TNF in Malarial Serum

Several laboratories have argued that TNF secretion occurs during malarial infections. In 1984, before the Rockefeller group was aware that cachectin and TNF were identical, they reported that lysed malaria parasites would trigger peritoneal macrophages to release cachectin-like activity.[71] Others have more recently shown that spleen cells[72] and serum[73] from malarial mice (*P. yoelii* and *P. berghei* ANKA, respectively) acquire cytotoxic activity that is abolished by antibody to recombinant mouse TNF. We have likewise assayed for TNF in plasma collected during the course of infection with *P. vinckei*, the parasite used in our studies on malarial pathology,[26-28] and found an exponential rise during the stage of the infection when tissue damage and biochemical changes begin (Figure 4).

Applying these techniques to human malaria is still in its infancy, but Scuderi et al.[74] recently reported seven out of ten human malarial sera from patients of undefined clinical status were positive for TNF. We are currently studying TNF concentrations in human malarial sera from southwest Pacific islands, and similar projects are being undertaken in Gambia and Geneva. It is important to appreciate in this context that elapsed time since the

lightly parasitized + TNF   terminally affected no TNF

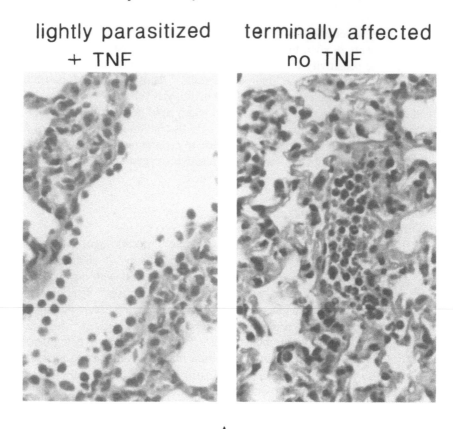

A

FIGURE 1 (A).   Neutrophil accumulations in mouse pulmonary venules, 6 h after a single (0.4 mg/kg) injection of TNF/cachectin into a lightly, *P. vinckei*-parasitized CBA mouse, and in an untreated, terminally-affected animal. Hematoxylin and eosin. (Original magnification × 400.) (From Clark et al., *Am. J. Pathol.*, 129, 192, 1987. With permission.) (B) Midzonal coagulative necrosis in mouse liver, 22 h after a single (0.4 mg/kg) injection of TNF/cachectin into a lightly *P. vinckei*-parasitized CBA mouse, and in an untreated, terminally affected animal. Hematoxylin and eosin. (Original magnification × 250.) (From Clark et al., *Am. J. Pathol.*, 129, 192, 1987. With permission.)

last wave of schizogony, rather than severity of illness at the time of sampling, is likely to determine how much TNF can be detected in an individual serum sample. Waage and co-workers, working on meningococcal septicemia, have routinely found the highest TNF values soon after admission of patients into the hospital, even though disease severity increased subsequently.[75,76] This is in keeping with the experience of ourselves[26] and others,[77] in which TNF-induced pathology occurs long after a bolus injection of TNF has cleared from the plasma.

The nature of the trigger for TNF release in malaria is also under investigation. The group from the Middlesex Hospital, London, has recently shown that mouse malarial parasites can *in vitro* serve as well as endotoxin in triggering TNF release.[78] In these experiments, polymyxin B was added to control for endotoxin contamination. An obvious approach for several groups will now be to see if the various defined malarial antigens that have been genetically engineered by laboratories attempting to make malarial vaccines will also trigger macrophages to release TNF.

The ultimate test for the *in vivo* importance of TNF in malarial pathology is whether injecting antibody directed against TNF will inhibit its development. This work has been retarded by the practical difficulty of obtaining enough recombinant mouse TNF to be able to raise antibodies against it. To date, only one experiment of this type has been reported

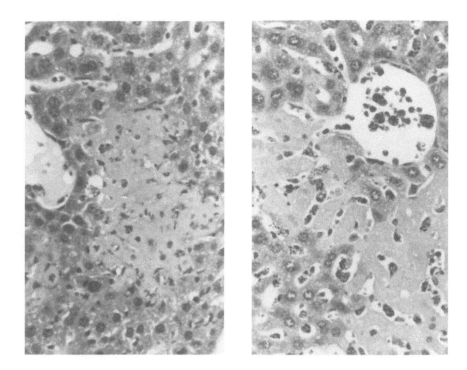

FIGURE B

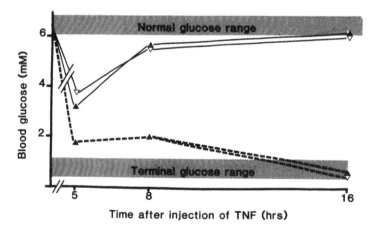

Time after injection of TNF (hrs)

FIGURE 2.   Time course of change in blood glucose level in normal or lightly (*P. vinckei*) parasitized CBA mice that received a single injection of TNF/ cachectin; (∇----∇) normal plus 0.4 mg/kg ($\blacktriangle$----$\blacktriangle$) normal plus 0.6 mg/kg; (∇———∇) parasitized plus 0.4 mg/kg; and ($\blacktriangle$———$\blacktriangle$) parasitized plus 0.6 mg/kg. Where larger than symbol size, standard errors of the mean are indicated. (From Clark et al., *Am. J. Pathol.*, 129, 192, 1987. With permission.)

from a group with an interest in the cerebral malaria caused in mice by *P. berghei* ANKA. They found that a single injection of antibody prevented the cerebral symptoms and pathology usually seen in this infection (Figure 5) and converted the disease into the type seen with other strains of *P. berghei*, with the mice living much longer and dying of other causes.

B. INTERACTION BETWEEN TNF AND OTHER MEDIATORS

It is becoming increasingly obvious that the actions of monokines and lymphokines

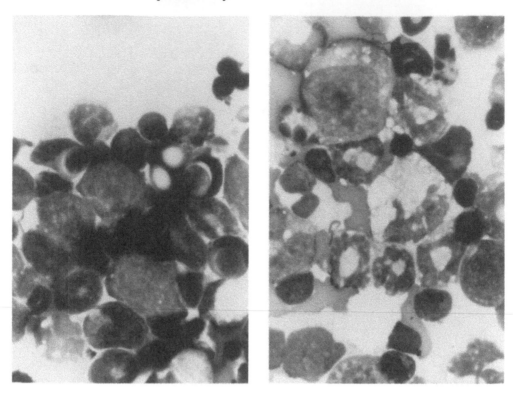

A   B

FIGURE 3 (A). Erythrophagocytosis in bone marrow 20 h after a CBA mouse with 5% parasitaemia of *P. vinckei* had received an injection of 5 μg r human TNF (Chiron). (B) Karyorrhexis of erythroid precursors produced under the same conditions. Both changes are also seen in terminal *P. vinckei* infections. (From Clark and Chaudhri, in *Tumor Necrosis Factor and Related Cytokines*. Heidelberg, Bonavida, B. and Kirchner, H., Eds., S. Karger, Basel, in press.)

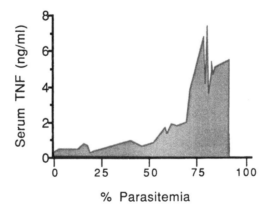

FIGURE 4. Serum TNF levels throughout the course of infection in individual *P. vinckei*-infected CBA mice. TNF was quantitated by a bioassay using actinomycin D sensitized WEHI-164 target cells. Recombinant human TNF from Chiron (5 × 10⁴ U/μg and 200 μg/ml) was used as control. The graph represents data from 32 mice with a range of parasitemias.

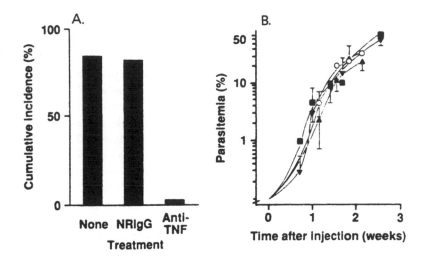

FIGURE 5.   Prevention of acute neurological complications in *P. berghei*-infected CBA/Ca mice treated with antibody to TNF-α. (A) Cumulative incidence of the neurological syndrome. (B) Parasitemia (bars indicate SD). Mice were injected intravenously with 2 mg of protein A (Pharmacia)-adsorbed rabbit antiserum to TNF-α or normal IgG in 0.5 ml of phosphate buffered saline, 0.01M, pH 7.2. This rabbit antibody was produced by immunization with purified, *Escherichia coli* recombinant mouse TNF-α in Freund's complete adjuvant. Seven mice were injected on day 7 (▲) and eight on day 4 (▼) with antibody to TNF-α. As a control, five mice were injected with normal rabbit IgG (NRIgG) on day 4 (■), and 14 mice received PbA alone (○). Similar treatment with IgG from rabbits injected only with Freund's complete adjuvant did not influence the course of cerebral malaria. (From Grau et al., *Science*, 237, 1210, 1987. With permission.)

depend on an interrelated network of enhancement, synergy, and inhibition. The high level of redundancy in the system implies that these responses are carefully orchestrated because they are essential to survival, even if in relatively unusual circumstances, such as acute infection, its components can lead to immunopathology through excessive production. Interactions that may be important in a malarial context are discussed here.

1. TNF and Products of Activated T Lymphocytes

Gamma-interferon (IFN-γ), a well-studied product of activated T cells, has been reported to be present in the serum of patients with falciparum malaria.[79] IFN-γ can increase both gene transcription for TNF and interleukin-1[80] and the expression of TNF receptors,[81] both of which probably help explain why it enhances the toxicity of TNF.[82] We have found that parenteral IFN-γ greatly enhances TNF production by *P. vinckei*-infected mice (Chaudhri and Clark, unpublished data), and that athymic mice, which lack the capacity to make IFN-γ, have much less tissue damage than intact controls when infected with this parasite.[83] There are also reports that splenomegaly, enhanced phagocytosis, and anemia in *P. yoelii*-infected mice are T cell-dependent[84] and that *P. berghei*-infected nude mice live longer than controls[85,86] with less cerebral pathology when the ANKA strain of parasite is used.[87] The capacity of TNF to cause erythrophagocytosis and dyserythropoiesis,[27] and its evident role in *P. berghei* ANKA-induced cerebral malaria,[73] strongly suggest that synergistic IFN-γ could be an important missing component of the response to these infections in nude mice. Interleukin-2, another product of activated T cells, can also enhance release of TNF *in vitro*[88] and *in vivo*.[89]

Conversely, TNF also influences the immunoregulatory function of IFN-γ in that it greatly increases the capacity of this lymphokine to induce expression of both class I and class II major histocompatability complex (MHC) antigens.[90] This may have implications for the autoimmune phenomena associated with malaria.

2. TNF and Interleukin-1

The term interleukin-1 was adopted in 1979 to rationalize the nomenclature of a monokine known even then to have various distinct functions.[91] It is now appreciated that at least two distinct polypeptide cytokines, termed IL-1α and IL-1β, produced by a diverse variety of cell types, have these properties and that, as reviewed recently,[38,92] the array of functions these cytokines possess is extraordinarily wide. These reviews also summarize the great functional overlap between IL-1 and TNF; both, for example, are endogenous pyrogens, and as such are equal candidates for generating the fever of malaria. These shared functions were initially puzzling, since comparisons of either form of IL-1 to TNF revealed no obvious homology in amino acid sequence. When secondary protein conformation was compared, however, close structural homology between these three molecules became apparent,[93] rationalizing their shared functions and suggesting their evolution from a common ancestral cytokine. Once this family tree is appreciated it is not so surprising that TNF induces release of IL-1 from endothelial cells[94] and monocytes,[95] and that the reverse induction evidently may occur.[96] Moreover, it is now (late 1987) beginning to emerge that TNF and IL-1 synergize powerfully.[97] When considered with their concommitant production under so many circumstances, this synergy implies that the experimental use of either one of these monokines alone, while instructive, gives only part of the true *in vivo* picture.

3. TNF and Prostaglandins

Although prostaglandins were first described over 50 years ago, it took 35 years for them to be regarded as mediators of inflammation.[98] Both prostaglandins and the leukotrienes have become prominent in the literature on inflammation (reviewed in Reference 99) and it is now appreciated that their main role in this context may be to enhance or otherwise regulate other mediators rather than to directly affect tissues. The altered body temperature, diarrhea, and hypoglycemia induced by TNF evidently involve prostaglandins or other products of the cyclooxygenase pathway, since these changes are inhibited by pretreatment with indomethacin and ibuprofen.[100] *In vitro*, indomethacin augments and exogenous PGE_2 inhibits LPS-induced TNF production,[101] while lower concentrations of PGE_2 paradoxically enhance *in vitro* TNF synthesis under these conditions.[102] Prostacyclin has also been shown to inhibit TNF production.[101] Thus, prostaglandins probably provide a self-regulatory mechanism for TNF formation since, like IL-1, TNF can induce prostaglandin production from a range of cells.[103,104] Once TNF is produced, however, exogenous PGE_2 (5 mg/kg) is reported to have no effect on a range of its biological activities.[101]

4. TNF and Reactive Oxygen Species

Several groups have demonstrated that TNF increases the readiness of leukocytes to release superoxide,[53,54] but nothing appears to have been published on whether oxidant stress on macrophages enhances their output of TNF. Encouraged by a report that radical scavengers inhibit lymphotoxin production by T lymphocytes,[105] we examined the effect of added oxidant stress (in the form of hydrogen peroxide or sodium periodate) on LPS-induced production by mouse peritoneal macrophages. We also investigated the effect of a range of free radical scavengers and iron chelators on TNF production under these conditions. Some of these results are shown in Figure 6. As shown, added oxidant stress enhanced TNF production and butylated hydroxyanisole or deferoxamine (desferrioxamine) inhibited its secretion. Other radical scavengers and iron chelators, as well as superoxide dismutase, tested produced similar results (not illustrated). Thus, mild oxidant stress may amplify TNF-induced tissue damage and metabolic change, a process to which superoxide secreted by leukocytes exposed to TNF could contribute. These events could help bring about the reported autocrine release of TNF from monocyte.[96] Our results also imply that TNF could act as an amplifying loop for oxidant damage to tissues and explain why either feeding diet high in radical-scavenging

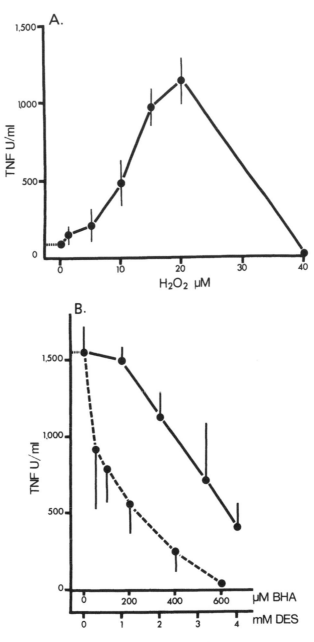

FIGURE 6. (A)   The effect of oxidant stress (H_2O_2) on TNF production. Peritoneal exudate cells were harvested from male CBA mice and cultured at 10^6 cells/ml in Dulbecco's Modified Eagle's medium supplemented with 10% heat inactivated fetal calf serum. These cells were suboptimally stimulated with 1 μg/ml lipopolysaccharide in the presence of various concentrations of H_2O_2, as indicated. The cultures were incubated at 37°C and 5% CO_2 atmosphere. Supernatants were then collected at 20 h and TNF quantitated by a bioassay using actinomycin D sensitized WEHI-164 target cells. r human TNF from Chiron (10^7 units/ml) was used as control. Data represents mean ± SEM of duplicates from three separate experiments. (From Clark and Chaudhri, *Iron, The Lymphoid System, Inflammation and Malignancy*, de Sousa, M. and Brock, J. Eds., John Wiley & Sons, in press. With permission.) (B) The effect of (●----●) butylated hydroxyanisole (BHA) and (●——●) desferrioxamine (DES) on TNF production. Peritoneal exudate cells were harvested and cultured as described in Figure 6A. These cells, in contrast, were optimally stimulated with 10 μg/ml lipopolysaccharide in the presence or absence of various concentrations of either BHA or DES, as indicated. Supernatants were collected at 20 h and TNF quantitated as for Figure 6A. Data represents mean ± SEM of duplicates from three separate experiments.

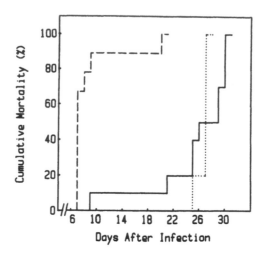

FIGURE 7.   Cumulative mortality of CBA mice infected with *P. berghei* ANKA. Mice were fed control diet (----) or food containing 0.75% BHA (——) or 0.5% ethoxyquin (···) from 7 d before i.p. inoculation with 10^6 parasitized erythrocytes on day 0 until death. (From Clark et al., in *Free Radicals, Oxidant Stress and Drug Action*, Rice-Evans, C., Ed., Richelieu Press, London, 237, 1987. With permission.)

nomenon of tolerance to TNF exists in man, since many groups report that tumor patients become refractory to TNF side effects on repeated exposure.[25] Tolerance to endotoxin is a well-established concept, and it has recently been reported that animals made tolerant to either endotoxin or TNF by repeated exposure are refractory to both agents.[23] Hence, it is very likely that the subjects in studies showing that untreated human malarial illness confers endotoxin tolerance[19,20] were also tolerant to TNF. As noted in Section II.A, even a single bout of untreated vivax malaria produces, on average, a fivefold tolerance of malaria parasites.[1] Thus, acquired malaria tolerance in human beings may be a manifestation of these recently recovered individuals being tolerant to TNF so that they need much more malaria-antioxidants[106] (Figure 7) or injecting antibody to TNF (Figure 8) are equally effective ways to prevent cerebral symptoms and pathology in *P. berghei* ANKA-infected mice.

5. TNF and Platelet-Activating Factor

As its trivial name implies, the phospholipid platelet-activating factor (PAF), or acetyl glyceryl ether phosphorylcholine, was discovered through its effects on platelets.[107] It was soon shown to be generated by leukocytes, vascular endothelial cells, and platelets, and proved, once synthetic preparations became available,[108] to have many diverse functions (reviewed in Reference 109). In brief, it causes hypotension associated with a large increase in plasma thromboxane,[10] activates neutrophils, aggregates platelets (and thus leads to a functional thrombocytopenia), and in certain circumstances increases vascular permeability. A recent review[111] covers its role in regulation of the cellular immune responses. Bessin and co-workers[112] were the first to notice parallels between PAF infusion and experimental endotoxicity, and others soon demonstrated that PAF antagonists would protect animals from i.v. endotoxin.[113-115] When considered alongside the association between endotoxicity and TNF,[67] these observations imply functional links between TNF and PAF and, not surprisingly, evidence has recently emerged that PAF enhances TNF production by LPS-treated monocytes and macrophages (submitted paper referred to in Reference 111) and that TNF is a very effective inducer of PAF synthesis and release from leukocytes and endothelial cells.[116] Through these steps PAF may, for instance, prove to have a key role in the hypotension, neutrophil adherence and priming, increased thromboxane, and thrombocytopenia

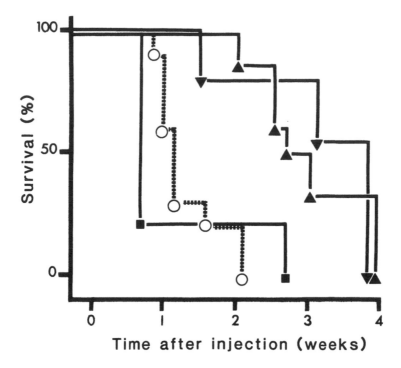

FIGURE 8. Cumulative mortality of *P. berghei*-infected CBA mice. Mice were treated with antibody to TNF-α as described in Figure 5. Seven mice were injected on day 7 (▲) and eight on day 4 (▼) with antibody to TNF-α. As a control, 5 mice were injected with normal rabbit IgG on day 4 (■), and 14 mice received PbA alone (○). (From Grau et al., *Science*, 237, 1210, 1987. With permission.)

now attributed to TNF. Thus, recognition of the importance of PAF as a mediator of inflammation continues to grow, and its association with TNF, which interacts with IFN-γ and IL-2, brings it into the realms of immunopathology.

IV. SOME POSSIBLE ROLES OF THESE MEDIATORS IN MALARIA

As we have noted, acute human malaria presents the experimental pathologist with a collection of symptoms and pathology that, given the intraerythrocytic location of the infectious agent, is not easily rationalized. The broader picture, which includes malarial tolerance and host species differences if even the same parasite species is involved (Section II), is still more complex, but must be accommodated in any proposal aspiring to give a coherent picture of the pathophysiology of the disease. In this last section of our chapter we give examples of how these different aspects of malaria, apparently so unconnected, are predictable consequences of the known interactions between the mediators of inflammation and host animals, including man.

As noted in Section III.A and reviewed in Reference 117, virtually all of the changes seen in acute human malaria have been reproduced by r TNF in humans (as side effects in TNF-treated tumor patients) or experimental animals. Furthermore, TNF is present in malarial sera. We now discuss other aspects of malaria, also requiring an explanation, in terms of TNF and the other cytokines with which it interacts.

A. MALARIAL TOLERANCE

As discussed in Section II.B.2, the innate tolerance to malaria and babesiosis, which determines the parasite density at which a previously unexposed animal will first become ill, correlates with that innate sensitivity of the species to endotoxin.[18] So far as has been examined, this correlates with sensitivity to exogenous TNF,[23,25,118] which is perhaps to be expected, given that TNF is an important mediator of endotoxicity.

We suggest that the acquired malarial tolerance that develops in untreated human malaria,[11,12] even after a single exposure, may also be understood in these terms. The phenomenon of tolerance to TNF exists in man, since many groups report that tumor patients become refractory to TNF side effects on repeated exposure.[25] Tolerance to endotoxin is a well-established concept, and it has recently been reported that animals made tolerant to either endotoxin or TNF by repeated exposure are refractory to both agents.[23] Hence, it is very likely that the subjects in studies showing that untreated human malarial illness confers endotoxin tolerance[19,20] were also tolerant to TNF. As noted in Section II.A, even a single bout of untreated vivax malaria produces, on average, a fivefold tolerance of malaria parasites.[1] Thus, acquired malaria tolerance in human beings may be a manifestation of these recently recovered individuals being tolerant to TNF so that they need much more malaria-induced TNF before they become ill again. Repeated exposure would enhance this state, and maintain it for some period, as is seen in malaria-tolerant children in hyperendemic areas.[11,12] This could be tested by assaying sequential plasma samples for TNF at schizogony and for several hours afterwards in malaria-tolerant individuals.

B. THE CONTRIBUTION OF SPLEEN CELLS TO MALARIAL PATHOLOGY

The possibility of a spleen-dependent immunopathological response in experimental malaria was raised when it was found that splenectomized mice lived longer and had less liver damage than controls when both groups were infected with *P. berghei*.[86,121] This was done in various strains of mice and could not be rationalized in terms of IgG antibodies.[119] A close parallel exists in *B. bovis*-infected cattle, where splenectomy reduces the reaction of the host to a given load of parasites.[120] It has been known for some time that splenectomy reduces the lethality of endotoxin both in normal mice[121] and those made more sensitive to endotoxin by prior injection of Bacillus Calmette Guerin (BCG).[122] Furthermore, the susceptibility of C3H/HeN mice to Westphal preparations of Gram-negative bacterial endotoxin, and the resistance of C3H/HeJ animals, have been shown by cell transfer experiments to reside largely in spleen cells.[123] There has, until now, been no rationale for these observations; in particular, altered clearance of endotoxin from the plasma could not account for the altered toxicity.[121] The firm links now made between endotoxin and TNF have concentrated efforts in this direction and it has recently been reported that splenectomized mice have significantly lower levels of TNF after injection of endotoxin.[124] We have confirmed this finding (Chaudhri and Clark, unpublished data). If this observation can be repeated in terminally ill malarial mice that have received no exogenous endotoxin it will provide a plausible explanation, open to further testing, for the central role of the spleen in malaria pathology.

C. CEREBRAL MALARIA

Cerebral malaria is the most important severe manifestion of *P. falciparum* infection in man and evidently occurs when, for some undetermined reason, parasitized red cells begin to favor the cerebral vascular bed as a site for sequestration, restricting blood flow.[125,126] It follows that understanding cerebral malaria largely depends on knowing what causes the onset of this predilection for the brain vasculature. It appears that red cells containing mature trophozoites and schizonts of *P. falciparum* usually adhere to endothelium, whether in Aotus monkeys,[13,16,17] malaria-tolerant children (not formally demonstrated, but can be assumed by the absence of these stages in the peripheral blood), or patients ill with falciparum malaria but not showing cerebral symptoms.[126] Thus, whether a patient with illness attributable to *P. falciparum* will develop cerebral symptoms (not due to hypoglycemia) is likely to be determined by an increased attractiveness of the endothelium of the cerebral vascular bed, rather than sequestration of red cells not previously adherent.

It is uncertain precisely what changes in the cerebral vascular endothelium bring this about. MacPherson et al.[126] have described endothelial projections, or pseudopodia, that they suggest could serve this purpose, but this explanation requires these pseudopodia to be a primary event, whereas they could, as in other circumstances,[127] be secondary to cerebral ischemia. Whether primary or secondary, however, these projections may be the work of monokines. TNF is circulating in severe malaria (Section III.2), and this monokine, in the presence of IFN-γ (reported present in falciparum malaria[79]) can induce cultured endothelial cells to extend processes.[128] It is conceivable that TNF could also play a role when these processes are secondary to cerebral ischemia,[127] since the oxidant stress of ischemia/reperfusion could help generate TNF, as discussed in Section III.B.4.

It appears that the receptor for cytoadherence of *P. falciparum*-infected red cells to endothelium is the glycoprotein thrombospondin.[129] Sources include endothelial cells,[130] and its synthesis can be induced by platelet-derived growth factor.[131,132] This, in turn, is induced in cultured endothelial cells by TNF[133] so TNF could prove to be an important determinant of how much thrombospondin is generated, and hence whether or not cytoadherence occurs. Thus, variation in thrombospondin concentrations on endothelial cells is plausibly an extension of systemic inflammation. Whether the endothelial cells of the cerebral vascular bed are more susceptible to thrombospondin induction is yet to be determined.

We have discussed the type of cerebral malaria seen in *P. berghei* ANKA-infected mice in Section III.B.4. Explaining this model presents a different set of problems, since adherence of knob-bearing red cells does not occur. Nevertheless, TNF appears to be involved in the pathogenesis of this condition since, as discussed, antibody to TNF prevents its occurrence (Figure 8).

V. CONCLUDING REMARKS

The recent flood of information on the polypeptide cytokines and their interactions with longer-established mediators of inflammation has broadened horizons in the pathophysiology of infectious diseases. Malaria is an excellent model on which to develop these concepts since it is clear, in a way not immediately obvious in the acute viral diseases, that the infectious agent can harm tissues it does not acutally invade. Within this network of cooperating lymphokines and monokines lie many opportunities for rational therapeutic intervention. Obvious examples are the use of neutralizing antibody directed against TNF or its receptors, and agents that inhibit its production or activity. However, the complexities of the patterns of interaction, and the range of cytokines involved, doubtlessly still hold many surprises. It is from these that the most useful therapies are likely to emerge.

ACKNOWLEDGMENTS

This study received support from the malarial component of the UNDP/World Bank/WHO Special Program for Research and Training in Tropical Disease and the National Health and Medical Research Council of Australia. We are grateful to Elizabeth Jackson for kindly typing the manuscript.

REFERENCES

1. **Kitchen, S. F.,** Symptomology: general considerations, in *Malariology,* Vol. 2, Boyd, M. R., Ed., W. B. Saunders, Philadelphia, 1949, 966.
2. **Kean, B. H. and Reilly, P. D.,** Malaria — the mime: recent lessons from a group of civilian travelers, *Am. J. Med.,* 61, 159, 1976.

3. **Hyman, A. S.**, Clinical masquerades of malaria, *U.S. Nav. Med. Bull.*, 45, 287, 1945.
4. **Anon.**, Severe and complicated malaria, a W.H.O. Malaria report, *Trans. R. Soc. Trop. Med. Hyg.*, 80 (Suppl. 1), 1986.
5. **Phillips, R. E. and Warrell, D. A.**, The pathophysiology of severe falciparum malaria, *Parasitol. Today*, 2, 271, 1986.
6. **Warrell, D. A.**, Pathophysiology of severe falciparum malaria in man, *Parasitology*, 94, 553, 1987.
7. **Golgi, C.**, Sull'infexione malarica, *Arch. Sci. Med.*, 10, 109, 1986.
8. **Young, M. D., Stubbs, T. H., and Coatney, G. R.**, Studies on induced quartian malaria in negro paretics, *Am. J. Hyg.*, C31, 51, 1940.
9. **Gaskell, J. F. and Millar, W. L.**, Studies on malignant malaria in Macedonia, *Q. J. Med.*, 13, 381, 1919.
10. **Meleney, H. E.**, The physiological pathology of malaria, in *A Symposium on Human Malaria*, American Association for Advanced Science, Washington, D.C., Publ. 15, 223, 1941.
11. **McGregor, I. A., Gilles, H. M., Walter, J. H., Davies, A. H., and Pearson, F. A.**, Effects of heavy and repeated malaria infections on Gambian infants and children, *Br. Med. J.*, 2, 686, 1956.
12. **Hill, R. B., Cambournac, F. J. C., and Simoes, M. P.**, Observations on the course of malaria in children in an endemic region, *Am. J. Trop. Med. Hyg.*, 23, 147, 1943.
13. **Jervis, H. R., Spring, H., Johnson, A. J., and Welde, B. T.**, Experimental infections with *Plasmodium falciparum* in Aotus monkeys. II. Observations on host pathology, *Am. J. Trop. Med. Hyg.*, 21, 272, 1972.
14. **Wellde, B. T., Johnson, A. J., Williams, J. S., and Sadun, E. H.**, Experimental infections with *Plasmodium falciparum* in Aotus monkeys. I. Parasitology, hematology and serum biochemical determinations, *Am. J. Trop. Med. Hyg.*, 21, 260, 1972.
15. **Schmidt, L. H.**, *Plasmodium falciparum* and *Plasmodium vivax* infections in the Owl monkey *(Aotus trivirgatus)*. The course of untreated infections, *Am. J. Trop. Med. Hyg.*, 27, 671, 1978.
16. **Voller, A., Hawkey, C. M., Richards, W. H. G., and Ridley, D. S.**, Human malaria *(Plasmodium falciparum)* in Owl monkeys *(Aotus trivirgatus)*, *J. Trop. Med. Hyg.*, 72, 153, 1969.
17. **Miller, L. H.**, Distribution of mature trophozoites and schizonts of *Plasmodium falciparum* in the organs of *Aotus trivirgatus*, the night monkey, *Am. J. Trop. Med. Hyg.*, 18, 860, 1969.
18. **Clark, I. A.**, Correlation between susceptibility to malaria and babesia parasites and to endotoxicity, *Trans. R. Soc. Trop. Med. Hyg.*, 76, 4, 1982.
19. **Heyman, A. and Beeson, P. B.**, Influence of various disease states upon the febrile response to intravenous injections of typhoid bacterial pyrogen, *J. Lab. Clin. Med.*, 34, 1400, 1949.
20. **Rubenstein, M., Mulholland, J. H., Jeffrey, G. M., and Wolff, S.**, Malaria-induced endotoxin tolerance, *Proc. Soc. Exp. Biol. Med.*, 118, 283, 1965.
21. **Healy, G. R. and Ruebush, T. K.**, Morphology of *Babesia microti* in human blood smears, *Am. J. Clin. Pathol.*, 73, 107, 1980.
22. **Kim, Y. B. and Watson, D. W.**, Role of antibodies in reactions to gram-negative bacterial endotoxins, *Ann. N. Y. Acad. Sci.*, 133, 727, 1966.
23. **Galanos, C. and Freudenberg, M.**, Tumor necrosis factor (TNF), a mediator of endotoxin lethality, *Immunobiology*, 175, 13, 1987.
24. **Saunter, C. and Wolfensberger, C.**, Interferon in human serum after injection of endotoxin, *Lancet*, ii, 852, 1980.
25. **Spriggs, D. R., Sherman, M. L., Frei, E., and Kufe, D. W.**, Clinical studies with tumor necrosis factor, in *Tumor Necrosis Factor and Related Cytotoxins*, Ciba Foundation Symposium No. 131, Bock, G. and Marsh, J., Eds. John Wiley & Sons, Chichester, England, 1987, 206.
26. **Clark, I. A., Cowden, W. B., Butcher, C. A., and Hunt, N. H.**, Possible roles of tumor necrosis factor in the pathology of malaria, *Am. J. Pathol.*, 129, 192, 1987.
27. **Clark, I. A. and Chaudhri, G.**, Tumour necrosis factor may contribute to the anemia of malaria by causing dyserythropoesis and erythrophagocytosis, *Br. J. Haematol.*, 70, 99, 1988.
28. **Clark, I. A. and Chaudhri, G.**, Tumour necrosis factor in malaria-induced abortion, *Am. J. Trop. Med. Hyg.*, 39, 246, 1988.
29. **Zuckerman, A. and Yoeli, M.**, Age and sex as factor influencing *Plasmodium berghei* infections in intact and splenectomized rats, *J. Infect. Dis.*, 94, 225, 1954.
30. **Tracey, K. J., Beutler, B., Lowry, S. F., Merryweather, J., Wolpe, S., Milsark, I. W., Hariri, R. J., Faheh, T. J., Zentella, A., Albert, J. D., Shires, G. T., and Cerami, A.**, Shock and tissue injury induced by recombinant human cachectin, *Science*, 234, 470, 1986.
31. **Berczi, I., Bertok, L., and Bereznai, T.**, Comparative studies on the toxicity of *Escherichia coli* lipopolysaccharide endotoxin in various animal species, *Can. J. Microbiol.*, 12, 1070, 1966.
32. **Zuckerman, A.**, *In vitro* opsoni tests with *Plasmodium gallinaceum* and *Plasmodium lophurae*, *J. Infect. Dis.*, 77, 28, 1980.
33. **Goodwin, M. H. and Stapleton, T. K.**, The course of natural and induced infections of *Plasmodium floridense* in *Sceloporus undulatus undulatus*, *Am. J. Trop. Med. Hyg.*, 1, 773, 1952.

34. **Rogers, R. N.,** Observations on the pathology of *Babesia argentine* infections in cattle, *Aust. Vet. J.,* 47, 242, 1971.

35. **Wright, I., Goodger, B. V., McKenna, R. V., and Mahoney, D. F.,** Acute *Babesia bovis* infection: a study of the vascular lesions in kidney and lung, *Z. Parasitekd.,* 60, 19, 1979.

36. **Maegraith, B.,** *Pathological Processes in Malaria and Blackwater Fever,* Blackwell Scientific, Oxford, 1948, 367.

37. **Larson, G. L. and Hensen, P. M.,** Mediators of inflammation, *Ann. Rev. Immunol.,* 1, 335, 1983.

38. **Moore, T. L. and Weiss, T. D.,** Mediators of inflammation, *Semin. Arthritis Rheum.,* 14, 247, 1985.

39. **Nathan, C. F.,** Secretory products of macrophages, *J. Clin. Invest.,* 79, 319, 1987.

40. **Gardner, S. M., Mack, B. A., Hilgers, J., Huppi, K. E., and Roeder, W. D.,** Mouse lymphotoxin and tumor necrosis factor: structural analysis of the cloned genes, physical linkage and chromosomal position, *J. Immunol.,* 139, 476, 1981.

41. **Clark, I. A., Wills, E. J., and Richmond, J. E.,** Intra-erythrocytic death of *Babesia* and *Plasmodium* during the immune response, *Trans. R. Soc. Trop. Med. Hyg.,* 10, 12, 1976.

42. **Taliaferro, W. H. and Taliaferro, L. G.,** Morphology, periodicity and course of infection of *Plasmodium brasilianum* in Panamanian monkeys, *Am. J. Hyg.,* 20, 1, 1934.

43. **Clark, I. A., Allison, A. C., and Cox, F. E. G.,** Protection of mice against *Babesia* and *Plasmodium* with BCG, *Nature,* 259, 309, 1976.

44. **Clark, I. A., Cox, F. E. G., and Allison, A. C.,** Protection of mice against *Babesia* spp. and *Plasmodium* spp. with killed *Corynebacterium parvum, Parasitology,* 74, 9, 1977.

45. **Clark, I. A.,** Resistance to *Babesia* spp. and *Plasmodium* spp. in mice pretreated with an extract of *Coxiella burneti, Infect, Immun.,* 24, 319, 1979.

46. **Clark, I. A.,** Does endotoxin cause both the disease and parasite death in acute malaria and babesiosis?, *Lancet,* 2, 75, 1978.

47. **Carswell, E. A., Old, L. J., Kassel, R. L., Green, S., Fiore, N., and Williamson, B.,** An endotoxin-induced serum factor that causes necrosis of tumors, *Proc. Natl. Acad. Sci. U.S.A.,* 72, 3666, 1975.

48. **Green, S., Dobrjansky, A., Chiasson, M. A., Carswell, E., Schwartz, M. K., and Lloyd, L. J.,** *Corynebacterium parvum* as the priming agent in the production of tumor necrosis factor in the mouse, *J. Natl. Cancer Inst.,* 59, 1519, 1977.

49. **Clark, I. A., Virelizier, J.-L., Carswell, E. A., and Wood, P. R.,** Possible importance of macrophage-derived mediators in acute malaria, *Infect. Immun.,* 32, 1058, 1981.

50. **Taverne, J., Tavernier, J., Fiers, W., and Playfair, J. H. L.,** Recombinant tumour necrosis factor inhibits malaria parasites *in vivo* but not *in vitro, Clin, Exp. Immunol.,* 67, 1, 1987.

51. **Clark, I. A., Hunt, N. H., Butcher, G. A., and Cowden, W. B.,** Inhibition of murine malaria (*Plasmodium chabaudi*) *in vivo* by recombinant interferon-γ or tumor necrosis factor, and its enhancement by butylated hydroxyanisole, *J. Immunol.,* 139, 3493, 1987.

52. **Jensen, J. B., Vande Waa, J. A., and Karadsheh, A. J.,** Tumor necrosis factor does not induce *Plasmodium falciparum* crisis forms, *Infect. Immun.,* 55, 1722, 1987.

53. **Klebanoff, S. J., Vadas, M. A., Harlan, J. M., Sparks, L. H., Gamble, J. R., Agosti, J. M., and Waltersdorph, A. M.,** Stimulation of neutrophils by tumor necrosis factor, *J. Immunol.,* 136, 4220, 1986.

54. **Shparber, M. and Nathan, C.,** Autocrine activation of macrophages by recombinant tumor necrosis factor but not recombinant interleukin-1, *Blood,* 68 (Suppl. 1), 86a, 1986.

55. **Cathcart, M. K., Morel, D. W., and Chisholm, G. M.,** Monocytes and neutrophils oxidise low density lipoprotein making it cytotoxic, *J. Leukocyte Biol.,* 38, 341, 1985.

56. **Buffinton, G. D., Hunt, N. H., Cowden, W. B., and Clark, I. A.,** Detection of short-chain carbonyl products of lipid peroxidation from malaria-parasite (*Plasmodium vinckei*)-infected red blood cells exposed to oxidative stress, *Biochem. J.,* 249, 63, 1988.

57. **Clark, I. A., Butcher, G. A., Buffinton, G. D., Hunt, N. H., and Cowden, W. B.,** Toxicity of certain products of lipid peroxidation to the human malaria parasite *Plasmodium falciparum, Biochem. Pharmacol.,* 36, 543, 1987.

58. **Clark, I. A.,** Suggested importance of monokines in pathophysiology of endotoxin shock and malaria, *Klin. Wochenschr.,* 60, 756, 1982.

59. **Richman, A. V., Gerber, L. I., and Balis, J. U.,** Peritubular capillaries, a major site of endotoxin-induced vascular injury in the primate kidney, *Lab. Invest.,* 43, 327, 1980.

60. **Moore, R. N., Goodrum, K. J., and Berry, L. J.,** Mediation of an endotoxic effect by macrophages, *J. Reticulo-Endothel. Soc.,* 19, 187, 1976.

61. **Buchanan, B. J. and Filkins, J. P.,** Insulin secretion and carbohydrate metabolic alterations and endotoxemia, *Circ. Shock,* 3, 267, 1976.

62. **Meyrick, B. and Brigham, K. L.,** Acute effects of *Escherichia coli* endotoxin on the pulmonary microcirculation of the anesthetized sheep, *Lab. Invest.,* 48, 458, 1983.

63. **Blackard, J. M., Hsieh, J., Fewel, J., and Rush, B. F.,** Tissue metabolites in endotoxin and hemorrhagic shock, *Arch. Surg. Chicago,* 107, 181, 1973.

64. **Kawakami, M. and Cerami, A.,** Studies on endotoxin-induced decrease in lipoprotein lipase activity, *J. Exp. Med.,* 134, 631, 1981.

65. **Weipers, W. L., Nagy, L., Pirie, H. M., Sanford, J., and Pillinger, A.,** Comparison of the toxic effects of intestinal obstruction fluid with those of certain endotoxins, *J. Pathol.,* 110, 295, 1973.

66. **Zahl, P. A. and Bjerkines, C.,** Induction of decidua-placental hemorrhage in mice by the endotoxins of certain gram-negative bacteria, *Proc. Soc. Exp. Med. Biol.,* 54, 329, 1943.

67. **Beutler, B., Milsark, I. W., and Cerami, A. C.,** Passive immunization against cachectin/tumor necrosis factor protects mice from lethal effect of endotoxin, *Science,* 229, 869, 1985.

68. **Ciancio, S. B., Jones, S. B., Yelich, M. R., and Filkins, J. P.,** Glucoregulatory and sympathoadrenal responses to TNF in conscious rats, *Fed. Proc. Fed. Am. Soc. Exp. Biol.,* 46, 561, 1987.

69. **Tracey, K. J., Beutler, B., Lowry, S. F., Merryweather, J., Wolpe, S., Milsark, I. W., Hariri, R. J., Fahey, T. J., Zentella, A., Albert, J. D., Shires, G. T., and Cerami, A.,** Shock and tissue injury induced by recombinant human cachectin, *Science,* 234, 470, 1986.

70. **Tracey, K. J., Lowry, S. F., Fahey, T. J., Albert, J. D., Fong, Y., Hesse, D., Beutler, B., Manogue, K. R., Calrano, S., Wei, H., Cerami, A., and Shires, G. T.,** Cachectin/tumor necrosis factor induces lethal shock and stress hormone response in the dog, *Surg. Gynecol. Obstet.,* 164, 415, 1987.

71. **Hotez, P. J., Le Trang, N., Fairlamb, A. H., and Cerami, A.,** Lipoprotein lipase suppression in 313-LI cells by a haemo protozoan-induced mediator from peritoneal exudate cells, *Para. Immunol.,* 6, 203, 1984.

72. **Taverne, J., Treagust, J. D., and Playfair, J. H. L.,** Macrophage cytotoxicity in lethal and non-lethal murine malaria and the effect of vaccination, *Clin. Exp. Immunol.,* 66, 45, 1986.

73. **Grau, G. E., Fajardo, L. F., Piquet, P.-F., Allet, B., Lambert, P.-H., and Vassali, P.,** Tumor necrosis factor (cachectin) as an essential mediator in murine cerebral malaria, *Science,* 237, 1210, 1987.

74. **Scuderi, P., Sterling, K. E., Lam, K. S., Finley, P. R., Ryan, K. J., Roy, C. G., Petersen, E., Shymen, D. J., and Salmon, S. E.,** Raised serum levels of tumor necrosis factor in parasitic infections, *Lancet,* ii, 1364, 1986.

75. **Waage, A., Espevik, T., Bakke, O., Halstensen, A., Brandtzaeg, P., Nissen-Meyer, J., and Lamvik, J.,** Detection of TNF in human septicaemia, *Immunobiology,* 175, 41, 1987.

76. **Waage, A., Brandtzaeg, P., Halstensen, A., Kierulf, P., and Espevik, T.,** The complete pattern of cytokines in serum from patients with meningococcal septic shock, *J. Exp. Med.,* 169, 333, 1989.

77. **Waage, A.,** Production and clearance of tumor necrosis factor in rats exposed to endotoxin and dexamethasone, *Clin. Immunol. Immunopathol.,* 45, 348, 1987.

78. **Bate, C. A. W., Taverne, J., and Playfair, J. H. L.,** Malaria parasites induce tumour necrosis factor production by macrophages, *J. Immunol.,* 64, 227, 1988.

79. **Rhodes-Feuillete, A., Bellosguado, M., Fruihle, P., Ballet, J. J., Chousterman, S., Canuvet, M., and Penes, P.,** The interferon compartment of the immune response in human malaria. II. Presence of serum interferon-gamma following the acute attack, *J. Interferon Res.,* 5, 169, 1985.

80. **Collart, M. A., Relin, D., Vassali, J.-D., DeKassado, S., and Varsalli, P.,** γ-Interferon enhances macrophage transcription of the tumor necrosis/cachectin, interleukin-1, and urokinase genes, which are controlled by short-lived suppressors, *J. Exp. Med.,* 164, 2113, 1986.

81. **Ruggerio, V., Tavernier, J., Fiers, W., and Baglioni, C.,** Induction of the synthesis of tumor necrosis factor receptors by interferon-γ, *J. Immunol.,* 136, 2445, 1986.

82. **Talmadge, J. E., Bowersox, O., Tribble, H., Lee, S. H., Shepard, M., and Liggitt, D.,** Toxicity of tumor necrosis factor is synergistic with γ-interferon and can be reduced with cyclooxygenase inhibitors, *Am. J. Pathol.,* 128, 410, 1987.

83. **Clark, I. A. and Clouston, W. M.,** Effects of endotoxin on the histology of intact and athymic mice infected with *Plasmodium vinckei* petteri, *J. Pathol.,* 131, 221, 1980.

84. **Roberts, D. W. and Weidanz, W. P.,** Splenomegaly, enhanced phagocytosis, and anemia are thymus-dependent responses to malaria, *Infect. Immun.,* 20, 728, 1978.

85. **Waki, S. and Suzuki, M.,** A study of malaria immunobiology using nude mice, in *Proceedings of the Second International Workshop on Nude Mice,* Gustav Fisher, Stuttgart, 1977, 37.

86. **Van Zon, A., Eling, W., and Jerusalem, C.,** Histo- and immunopathology of malaria (*Plasmodium berghei*) infection in mice, *Br. J. Med.,* 14, 659, 1978.

87. **Finley, R. W., Mackey, L. J., and Lambert, P. H.,** Virulent *P. berghei* malaria: prolonged survival and decreased cerebral pathology in T cell-deficient nude mice, *J. Immunol.,* 129, 2213, 1982.

88. **Nedwin, G. E., Svedersky, L. P., Bringmann, T. S., Palladino, M. A., and Goeddel, D. V.,** Effect of interleukin-2, interferon-γ and mitogens on the production of tumor necrosis factors α and β, *J. Immunol.,* 135, 2492, 1985.

89. **Palladino, M. A.,** In discussion, p. 18, of Gifford, G. E., and Flick, D. A., Natural production and release of tumour necrosis factor, in *Tumour Necrosis Factor and Related Cytotoxins,* Ciba Foundation Symposium No. 131, Bock, G. and Marsh, J., Eds., John Wiley & Sons, Chichester, England, 1987, 3.

90. **Pfizenmaier, K., Scheurich, P., Schluter, C., and Kronke, M.,** Tumor necrosis factor enhances HLA-A, B, C and HLA-DR gene expression in human tumor cells, *J. Immunol.,* 138, 975, 1987.

91. **Aarden, L. A., Brunner, T. K., Cerottini, J.-C., et al.,** Revised nomenclature for antigen-nonspecific T cell proliferation and helper functions, *J. Immunol.,* 123, 2928, 1979.

92. **Le, J. and Vilcek, J.,** Biology of disease: tumor necrosis factor and interleukin-1: cytokines with multiple overlapping biological activities, *Lab. Invest.,* 56, 234, 1987.

93. **Cohen, F. E. and Dinarello, C. A.,** Structural homology between interleukin-1 and tumor necrosis factor, *J. Leuk. Biol.,* 42, 548, 1987.

94. **Nawroth, P. P., Bank, I., Handley, D., Cassimeris, J., Chess, L., and Stern, D.,** Tumor necrosis factor/cachectin interacts with endothelial cell receptors to induce release of interleukin-1, *J. Exp. Med.,* 163, 1363, 1986.

95. **Dinarello, C. A., Cannon, J. G., Wolff, S. M., Bernheim, H. A., Beutler, B., Cerami, A., Figari, I. S., Palladino, M. A., and O'Connor, J. V.,** Tumor necrosis factor (cachectin) is an endogenous pyrogen and induces production of interleukin 1, *J. Exp. Med.,* 163, 1433, 1986.

96. **Philip, R. and Epstein, L. B.,** Tumour necrosis factor as immunomodulator and mediator of monocyte cytotoxicity induced by itself, gamma-interferon and interleukin-1, *Nature,* 323, 86, 1986.

97. **Gowen, M.,** Actions of IL-1 and TNF on osteoblast-like cells: similarities and synergism, *J. Leukocyte Biol.,* 42, 546, 1987.

98. **Willis, A. L.,** Identification of prostaglandin E_2 in the rat inflammatory exudate, *Pharmacol. Res. Commun.,* 2, 297, 1970.

99. **Davies, P., Bailey, P. J., Goldenberg, M. M., and Ford-Hutchinson, A. W.,** The role of arachidonic acid oxygenation products in pain and inflammation, *Annu. Rev. Immunol.,* 2, 335, 1984.

100. **Kettlehut, I. C., Fiers, W., and Goldberg, A. L.,** The toxic effects of tumor necrosis factor *in vivo* and their prevention by cyclooxygenase inhibitors, *Proc. Natl. Acad. Sci. U.S.A.,* 84, 4273, 1987.

101. **Kunkel, S. L., Larrick, J. W., and Remick, D. G.,** Prostaglandin E_2 (PGE$_2$) and prostacyclin (PGI$_2$) regulate tumor necrosis factor-α (TNF) production at the cellular and molecular level; an analysis of autocrine, paracrine, and endocrine effects, *Immunobiology,* 175, 80, 1987.

102. **Renz, H., Nain, M., Gong, J.-H., and Gemsa, D.,** Prostaglandin E_2 (PGE$_2$) dose dependently stimulates or suppresses tumor necrosis factor-α (TNF) production, *Immunobiology,* 175, 135, 1987.

103. **Mizel, S. B., Dayer, J. M., Krane, S. M., and Mergenhagen, S. E.,** Stimulation of rheumatoid synovial cell collagenase and prostaglandin production by partially purified lymphocyte-activating factor (interleukin-1), *Proc. Natl. Acad. Sci. U.S.A.,* 78, 2474, 1981.

104. **Dayer, J. M., Beutler, B., and Cerami, A.,** Cachectin/tumor necrosis factor stimulates collagenase and prostaglandin E_2 production by human synovial cells and dermal fibroblasts, *J. Exp. Med.,* 162, 2163, 1985.

105. **Kobayashi, Y., Asada, M., and Osawa, T.,** Mechanism of phorbol myristate acetate-induced lymphotoxin production by a human T cell hybridoma, *J. Biochem.,* 95, 1775, 1984.

106. **Clark, I. A., Chaudhri, G., Thumwood, C. M., Hunt, N. H., and Cowden, W. B.,** Free radicals in malaria immunopathology, in *Free Radicals, Oxidant Stress and Drug Action,* Rice-Evans, C., Ed., Richelieu Press, London, 1987, 237.

107. **Benveniste, J., Henson, P. M., and Cochrane, G. G.,** Leukocyte-dependent histamine release from rabbit platelets: the role of IgE, basophils, and a platelet-activating factor, *J. Exp. Med.,* 136, 1356, 1972.

108. **Godfroid, J. T., Heymans, F., Michel, E., Redeuilh, C., Steiner, C., and Benveniste, J.,** Platelet-activating factor (PAF-acether): total synthesis of 1-*O*-octacetyl 2-*O*-acetyl *sn*-glycero-3-phosphorylcholine, *FEBS Lett.,* 116, 161, 1980.

109. **Braquet, P., Touqui, L., Shen, T. Y., and Vargaftig, B. B.,** Perspectives in platelet-activating factor research, *Pharmacol. Rev.,* 39, 97, 1987.

110. **Lefer, A. M., Muller, H. F., and Smith, J. B.,** Pathophysiological mechanisms of sudden death induced by platelet-activating factor, *Br. J. Pharmacol.,* 83, 125, 1984.

111. **Braquet, P. and Rola-Pleszczynski, M.,** Platelet-activing factor and cellular immune responses, *Immunol. Today,* 8, 345, 1987.

112. **Bessin, P., Bonnet, J., Apffel, P., Soulard, C., Desgrou, L., Pelassi, I., and Benveniste, J.,** Acute circulatory shock caused by platelet-activating factor (PAF-acether) in dogs, *Eur. J. Pharmacol.,* 86, 403, 1983.

113. **Terashita, Z., Imura, Y., Nishikawe, K., and Sumida, S.,** Is platelet-activating factor (PAF) a mediator of endotoxin shock?, *Eur. J. Pharmacol.,* 109, 257, 1985.

114. **Doebber, T. W., Wu, M. S., Robbins, J. C., Choy, B. M., Chang, M. N., and Shen, T. Y.,** Platelet activating factor (PAF) involvement in endotoxin-induced hypotension in rats. Studies with PAF-receptor antagonist kadsurenone, *Biochem. Biophys. Res. Commun.,* 29, 799, 1985.

115. **Cosals-Stenzel, J.,** Protective effect of WEB 2086, a novel antagonist of platelet-activating factor, in endotoxin shock, *Eur. J. Pharmacol.,* 135, 117, 1987.

116. **Camussi, G., Bussolino, F., Salvidio, G., and Baglioni, C.,** Tumor necrosis factor/cachectin stimulates peritoneal macrophages, polymorphonuclear neutrophils and vascular endothelial cells to synthesise and release platelet-activating factor, *J. Exp. Med.,* 166, 1390, 1987.

117. **Clark, I. A.**, Cell-mediated immunity in protection and pathology of malaria, *Parasitol. Today*, 3, 300, 1987.
118. **Cerami, A. and Old, L. J.**, In discussion, p. 35, of Palladino, M. A., Patton, J. S., Figari, I. S., and Shalaby, M. R., Possible relationships between *in vivo* antitumour activity and toxicity of tumour necrosis factor-α, in *Tumour Necrosis Factor and Related Cytotoxins*, Ciba Foundation Symposium No 131, Bock, G. and Marsh, J., Eds., John Wiley & Sons, Chichester, England, 1987, 21.
119. **Eling, W. M. C.**, Role of spleen in morbidity and mortality of *Plasmodium berghei* infection in mice, *Infect. Immun.*, 30, 635, 1980.
120. **Wright, I. G. and Goodger, B. V.**, Pathogenesis of babesiosis, in *Babesiosis of Domestic Animals and Man*, Ristic, M., Ed., CRC Press, Boca Raton, FL, 1988, 99.
121. **Agarwal, M. K., Parant, M., and Parant, F.**, Role of spleen in endotoxin poisoning and reticuloendothelial function, *Br. J. Exp. Med.*, 53, 485, 1972.
122. **Chedid, L., Rousselot, C., and Parant, F.**, *In vitro* fixation and degradation of radioactive endotoxin by the RES of BCG-treated mice, *Adv. Exp. Med. Biol.*, 15, 173, 1971.
123. **Glode, M. L., Mergenhagan, S. E., and Rosenstreich, D. L.**, Significant contribution of spleen cells in mediating the lethal effect of endotoxin *in vivo*, *Infect. Immun.*, 14, 626, 1976.
124. **Fung, K. P., Choy, Y. M., and Lee, C. Y.**, Effects of anti-inflammatory drugs, mannoheptulose and splenectomy on lipopolysaccharide-induced tumour necrosis factor release and animal mortality, *Immunobiology*, 175, 12, 1987.
125. **Raventos-Suarez, C., Kaul, D. K., Macaluso, F., and Nagel, R. L.**, Membrane knobs are required for the micro-circulatory obstruction induced by *Plasmodium falciparum*-infected erythrocytes, *Proc. Natl. Acad. Sci. U.S.A.*, 82, 3829, 1985.
126. **Macpherson, G. G., Warrell, M. J., White, N. J., Looareesunan, S., and Warrell, D. A.**, Human cerebral malaria: a quantitative ultrastructural analysis of parasitized erythrocyte sequestration, *Am. J. Pathol.*, 119, 385, 1985.
127. **Dietrich, W. D., Busto, R., and Ginsberg, M. D.**, Cerebral endothelial microvilli: formation following global forebrain ischemia, *J. Neuropathol. Exp. Neurol.*, 43, 72, 1984.
128. **Stolpen, A. H., Guinan, E. C., Fiers, W., and Pober, J. S.**, Recombinant tumour necrosis factor and immune interferon act singly and in combination to reorganise human vascular endothelial cell monolayers, *Am. J. Pathol.*, 123, 16, 1986.
129. **Roberts, D. D., Sherwood, J. A., Spitalnik, S. L., Panton, L. J., Howard, R. J., Dixit, V. M., Frazier, W. A., Miller, L. H., and Ginsburg, V.**, Thrombospondin binds falciparum malaria parasitized erythrocytes and may mediate cytoadherence, *Nature*, 318, 64, 1985.
130. **Mosher, D. F., Doyle, M. J., and Jaffe, E. A.**, Synthesis and secretion of thrombospondin by cultured human endothelial cells, *J. Cell. Biol.*, 93, 343, 1982.
131. **Majack, R. A., Cook, S. C., and Bornstein, P.**, Platelet-derived growth factor and heparin-like glycosaminoglycans regulate thrombospondin synthesis and deposition in the matrix by smooth muscle cells, *J. Cell. Biol.*, 101, 1059, 1985.
132. **Asch, A. S., Lueng, L. L. K., Shapiro, J., and Nachman, R. L.**, Human brain giant cells synthesise thrombospondin, *Proc. Natl. Acad. Sci. U.S.A.*, 83, 2904, 1986.
133. **Hajjar, K. A., Hajjar, D. P., Silverstein, R. L., and Nochman, R. L.**, Tumor necrosis factor-mediated release of platelet-derived growth factor from cultured endothelial cells, *J. Exp. Med.*, 166, 235, 1987.

Chapter 7

GENETIC CONTROL OF HOST RESISTANCE TO MALARIA

Mary M. Stevenson

TABLE OF CONTENTS

I. INTRODUCTION

Entry into a host of parasites of the malaria-causing *Plasmodium* species and their subsequent intraerythrocytic multiplication presents the individual with a formidable foe. Presented with a vast array of antigens which can vary depending upon the stage of infection, the host can respond to the initial and ensuing infection in several ways. This can occur via several unique mechanisms including (1) innate or natural resistance mechanisms which can involve either the absence of membrane receptors, structural, or biochemical abnormalities of the erythrocyte or, as yet undefined, innate or natural immune mechanisms, (2) humoral immune mechanisms, or (3) cell-mediated immune mechanisms. Furthermore, the interplay between these mechanisms and the genetic background of the host adds another dimension to our attempts to understand the nature of host defense against malaria.

The relationship between heredity and resistance to infectious diseases has long been observed in a variety of experimental, as well as clinical, situations including malaria. The genetic background of the host represents perhaps one of the strongest influences on the magnitude of host responses to invading microorganisms. In some instances, it determines the outcome of the infection; that is, whether the host or the pathogen will be the victor. Still another factor, whose importance has only more recently been considered largely due to advances in molecular genetics, is the interplay between parasite and host genes. The genetic diversity existing among strains of a given species of *Plasmodium* adds further complexity to the relationship between the malaria-causing parasite and its respective host.

This chapter will review what is known about genetic control of host resistance to malaria in humans and in murine models, its implications on the pathological manifestations of the disease, and its influence on the successful vaccination of individuals in high-risk areas. The information to be discussed will include observations made in experimental murine models using the rodent malaria species *P. berghei, P. yoelii,* and *P. chabaudi.* As well, observations from clinical situations involving the human malaria species, *P. falciparum* and *P. vivax,* will be discussed.

II. GENETIC CONTROL OF INNATE RESISTANCE TO MALARIA IN MAN

Initiation of the asexual erythrocytic cycle of malaria requires attachment of the merozoites to a specific membrane receptor, followed by invasion and the successful multiplication of the parasite inside the red cell of the host. Prevention of any one of these steps can result in resistance to malaria. As will be discussed in this section, inheritance of any one of several specific erythrocyte-associated aberrations may result in prevention of one of these steps and thus, innate resistance to malaria (Table 1).

A. MEMBRANE RECEPTORS

It has long been known that there is strict species specificity between the host and the *Plasmodium* parasite. This is due to the requirement for parasite-specific receptors. The first demonstration of the identity of one such membrane receptor required for infection with *P. vivax* was the observation by Miller et al.[1] that individuals whose erythrocytes are Duffy blood group negative (i.e., those which lack the blood group determinants, Fy^a and Fy^b, designated genotypically as *FyFy*) are resistant to infection with *P. vivax*. The basis of this observation was derived from the long-standing observation of an association between the high frequency of the Duffy negative genotype and resistance to *P. vivax* in Africa.[2,3] Direct experimental evidence derived from *in vitro* studies that the Duffy blood group factor had a role in infection of erythrocytes by *P. vivax* merozoites was the finding that enzymatic removal of the determinant or treatment with anti-Fy^a antibody blocked invasion of the

TABLE 1
Major Genetically Determined Traits Conferring Innate Resistance to Malaria in Humans

Trait

Absence of Duffy blood group antigen	Involved in invasion of erythrocytes by *P. vivax*
En(a−) erythrocytes	Reduced invasion by *P. falciparum* due to lack of glycophorin A
Ovalocytosis	Altered cytoskeleton resulting in decreased invasion by *P. vivax*, *P. malariae* and lower parasite densities of *P. falciparum*
Presence of sickle cell hemoglobin or hemoglobin S in homozygous or heterozygous form	Hemoglobin S limits intraerythrocytic development of *P. falciparum* under conditions of oxidant stress *in vitro* and probably in tissues *in vivo*
α- or β-thalassemia	Abnormal hemoglobin results in defective intraerythrocytic development of parasites
Glucose-6-phosphate dehydrogenase deficiency	Decreased intraerythrocytic development of *P. falciparum* particularly under conditions of oxidant stress

treated erythrocytes by *P. knowlesi*, a simian parasite that can infect man and can invade Duffy blood group positive erythrocytes *in vitro*.[4] This finding correlated with the observation that Duffy negative erythrocytes were resistant to invasion *in vitro* with *P. knowlesi*.[4] These studies, however, further demonstrated that attachment of *P. knowlesi* merozoites occurred equally well in Duffy positive and negative cells. Invasion occurred only in Duffy positive erythrocytes. These investigators concluded that the Duffy factor is involved in the attachment of *P. vivax* merozoites to the erythrocyte membrane while another receptor, which appears to be associated with the Duffy factor, is involved in invasion.[4]

Successful invasion of erythrocytes by merozoites of *P. falciparum*, the major causative *Plasmodium* species of human malaria, appears to require a distinct receptor. During analysis of human red cells lacking various blood group antigens, Miller et al.[5,6] observed that En(a−)cells which lack the major transmembrane sialyglycoprotein, glycophorin A, had reduced invasion by *P. falciparum* but were invaded normally by *P. knowlesi*. This observation has now been confirmed in a number of other laboratories and a key role for glyclophorin A in the attachment of the parasite to the cell membrane has been suggested.[7,8] En(a−) cells, although relatively resistant to invasion by *P. falciparum* merozoites, can support parasite development if they are invaded. There is presently no information available on either the prevalence of En(a−) erythrocytes among inhabitants of high-risk malaria areas and, thus, the selective advantage of this trait for malaria resistance.

B. OTHER MEMBRANE ALTERATIONS

Another example of the relationship between human erythrocyte variants and susceptibility to malaria infection is the observation that up to 20% of Melanesians in areas of Papua New Guinea which are epidemic for malaria have decreased rates of infection with *P. vivax* and *P. malariae* and lower parasite densities of *P. falciparum*.[9] The red cells of these individuals, called ovalocytes because of their characteristic shape, have depressed expression of some blood group antigens, including I^T, I^F, LW, D, C, e, S, s, u, Kp^b, Jk^a, Xg^a, Scl, En^a, and Wr^b, and appear to have increased thermal stability.[10,11] The resistance of these cells to malarial invasion, however, appears to be due to an altered cytoskeletal structure rather than the alteration or absence of a specific erythrocyte receptor.[11-13] Since ovalocytosis in Papua New Guinea is genetically determined, correlation between the high geographical incidence of the trait and endemic malaria suggests that ovalocytosis confers a selective advantage against malaria.

C. RELATIONSHIP BETWEEN HEMOGLOBINOPATHIES AND ENZYME DEFICIENCIES OF ERYTHROCYTES AND RESISTANCE TO MALARIA

Because the intraerythrocytic multiplication of parasites of the *Plasmodium* species results in the symptomatology and pathology associated with malaria, inability of the parasite to develop normally once it invades the cell will result in less severe infections. In 1954, Allison[14] suggested an association between the high incidence of the trait for sickle cell hemoglobin and resistance to falciparum malaria in areas of Africa where this form of malaria is epidemic. Later studies showed that individuals with the sickle cell trait, either in the homozygous or heterozygous form (i.e., genotypically either SS or AS), and those with the trait of normal hemoglobin (AA) had the same incidence of infection but the severity of the disease differed.[15,16] The basis of less severe malaria in individuals with SS or AS hemoglobin appears to be due to the inability of the parasite to develop normally at low oxygen tensions which cause the cell to sickle.[17] It has also been observed that invasion of cells containing hemoglobin S is prevented under low oxygen tensions;[18] thus, the presence of hemoglobin S appears to limit both the invasion by *P. falciparum* and its development within erythrocytes under conditions of reduced oxygen to which parasitized red cells are exposed, particularly in the spleen.[19]

In addition to hemoglobin S, the frequency of two other abnormal hemoglobins, C and E, appears to be high in areas of endemic malaria.[20,21] Other genetically determined hemoglobinopathies which appear to have a protective effect against malaria include β-thalassemia in which there is a deficiency in synthesis of the β-chain of hemoglobin, due to several distinct mutations,[19,22] and α-thalassemia, due to deletion of either one or both genes encoding α-globin.[23] Similar to sickle cell anemia, the gene frequencies for β- and α-thalassemia correlate with malaria endemicity. Similarly, the presence of fetal hemoglobin, which persists for a time after birth and is found in some red cells of adults, results in resistance to malaria.[24] The persistance of fetal hemoglobin may partially account for the observation that newborns are protected against malaria.

Another genetically determined erythrocyte abnormality whose distribution occurs frequently in malarious areas of Africa is glucose-6-phosphate dehydrogenase (G6PD) deficiency. The distribution of this trait was first described by Allison and Clyde.[25,26] Heterozygote carriers of the deficiency appear to have less severe infections although this point was controversial in the past.[20] However, more recent studies performed using *in vitro* cultures of *P. falciparum* have demonstrated that parasites develop less well in G6PD-deficient erythrocytes under normal oxygen tension as well as under conditions of oxidant stress.[27,28] G6PD functions together with several other enzymes, including glutathione peroxidase and glutathione reductase, to provide protection of the erythrocyte against the oxidants, O_2^- and H_2O_2. *Plasmodium* parasites themselves can generate these oxidants within the infected red cell.[29] This contributes another source of oxidant stress on the erythrocyte. Further studies have indicated that there is excess release of ferriheme, due to denaturation of hemoglobin in G6PD-deficient cells, which can mediate inhibition of parasite growth.[30] Ferriheme release in response to the oxidant stress imposed by the malaria parasites inside the erythrocyte may account for the genetic advantage of G6PD deficiency in malarious areas.

III. GENETIC CONTROL OF SPECIFIC IMMUNITY TO MALARIA IN MAN

A. ANTIMALARIAL ANTIBODY TITERS AND THE MAJOR HISTOCOMPATIBILITY COMPLEX

Studies conducted among human populations of malarious areas of Africa have suggested a role for the MHC in determining levels of antimalarial antibodies.[31] Individuals having a

TABLE 2
Genetic Control of Resistance to Various
Plasmodium Species and Strains Among Inbred,
Congenic and Recombinant Inbred Mouse Strains

Murine *Plasmodium* Species and Strain	Ref.
Plasmodium berghei kasapa (K173)	41—51
P. berghei ANKA	52, 53, 70, 71,73
P. yoelii 17XNL	54—57
P. yoelii 17XL	57
P. chabaudi AS	58—64, 68
P. chabaudi adami	54

combination of two class I antigens, A2 and AW30, were found to have high titers of antibody to falciparum antigens.

Since T cells are involved as helper cells for antibody production during malaria, a role for immune response (IR) genes, which map to the MHC, is implicated in immune responses to malaria. Recent studies (to be discussed in Section VI) using H-2 congenic mouse strains have demonstrated that immune responses to recombinant and synthetic antisporozoite vaccines as well as to the entire circumsporozoite protein of *P. falciparum* are under Ir gene control.[32,33] The authors of these studies suggest that if an analogous situation exists in humans, the development of an effective vaccine will be hampered. In retrospect, the relationship between resistance to malaria and the human MHC needs to be reexamined.

IV. GENETIC CONTROL OF RESISTANCE TO MALARIA IN MICE

The genetic basis of host resistance to infection with three of the four rodent malaria species has now been established. Results from various laboratories have demonstrated genetic control of resistance to various strains of *P. berghei*, *P. yoelii*, and *P. chabaudi* (Table 2). In general, analysis of the level of resistance to malaria among various inbred strains and F1 and F2 hybrid and backcross progeny derived from representative resistant and susceptible parental strains has been utilized. H-2 congenic mouse strains have been used to establish the role of the MHC in resistance. More recently, analysis of the level of resistance of recombinant inbred (RI) strains derived by inbreeding of the F2 generation of a resistant and a susceptible progenitor has been used. Since data obtained with RI strains is cumulative, RI strains provide not only a useful tool to determine if the trait of resistance to a particular *Plasmodium* species is controlled by a single gene or multiple genes but serve two other important purposes. First, establishment of the strain distribution pattern (SDP) of inheritance of resistance to a particular malaria species allows for comparison with the SDP of other traits, thus, establishing linkage. Such information may be useful to determine if the mechanisms of resistance to two different *Plasmodium* species are similar or to ascertain the resistance mechanism if the SDP is identical to that of a known trait. Second, concordance of the SDP of malaria resistance with that of a trait(s) whose chromosomal location is known will establish the chromosomal location of the resistance gene or genes. Several important similarities and differences between the rodent malaria species must be pointed out since these may account for the observed similarities and differences among strains of mice in genetic control of resistance to malaria. In general, *P. berghei* and *P. yoelii* infect reticulocytes while *P. chabaudi* and *P. vinckei* preferentially invade mature erythrocytes.[34,35] The mechanism of control and elimination of infection with one strain of *P. yoelii*, *P. yoelii* 17 XNL, has been demonstrated to occur via production of a T cell-dependent antibody while that of at least two strains of *P. chabaudi* and of one strain of *P. vinckei* occurs via antibody-independent, T cell-dependent immune responses[36-39] (see also Chapter 2). Furthermore, immunity resulting from infection with a given murine *Plasmodium* species is of limited specificity.[39,40] For example, immunization with *P. chabaudi adami* will protect against reinfection with the homologous parasite or the normally lethal *P. vinckei* but will not cross protect against *P. berghei* or *P. yoelii*

A. GENETIC CONTROL OF RESISTANCE TO *PLASMODIUM BERGHEI*

Differences among inbred strains of mice in response to malaria were first described by Greenberg et al.[41-49] in a series of papers published in the 1950s using the Kasapa strain or K173 strain of *P. berghei*. In their initial studies, these investigators used survival as the criterion for resistance or susceptibility. They examined 12 inbred strains and 13 F1 hybrid combinations and found that the inbred strains fell into a short, an intermediate, or a long survival group while the F1 hybrids, in general, survived longer than the longest surviving parent. Among the inbred strains, C57BL and C57L mice were found to be resistant and DBA/2 and A/LN mice were found to be susceptible. Female mice were found to survive significantly longer than male mice of the same strain and age. Determination of the time to peak parasitemia showed a significant difference between a susceptible strain, such as DBA/2, and a resistant strain, such as C57BL. When survival was used as the marker of resistance, results of analysis of F1 hybrid, backcross, and F2 populations derived from resistant C57BL and susceptible DBA/2 parental strains were consistent with the hypothesis that resistance to *P. berghei* K173 in this strain combination is genetically controlled by a single, dominant, autosomal gene or closely linked set of genes. Mendalian analysis of progeny derived from a cross between outbred Swiss mice and an inbred strain suggested that several genes controlled resistance measured in terms of survival in this strain combination.

The difficulty with this model is that all strains of mice eventually succumb to infection. The strain-specific features of *P. berghei* K173 infection were, however, examined in six inbred strains of mice by studying parasitemia, weight changes in body, thymus, spleen and liver, serum glutamic oxalacetate transaminase activity, and histological changes in the spleen, liver, and thymus.[50,51] The results of these studies demonstrated that marked changes occurred in each of the parameters studied which varied according to the mouse strain examined. No one parameter was of prognostic value in predicting whether the strain would exhibit short- or long-term survival.

Variation in the level of resistance to infection with another strain of *P. berghei*, *P. berghei* ANKA, has also been demonstrated.[52,53] Analysis of the level of resistance in four inbred strains and one outbred strain showed that the level of parasitemia, the degree of anemia, and the level of specific antibody, although characteristic for each strain, did not correlate with susceptibility. Although immune complexes containing either IgM or IgG occurred in all strains and the kinetics of their appearance and their levels varied according to the mouse strain, the amount and type of immune complexes were not related to susceptibility. Histological examination of brain, kidney, liver, and spleen specimens from the five mouse strains demonstrated that, while histological changes and immunopathology were apparent in the tissue from all the strains, there were marked cerebral lesions only in the susceptible strains. The investigators concluded that cerebral lesions play a significant role in the etiology of acute death in this murine malaria system which may serve as a model of human cerebral malaria. Information concerning the number of genes controlling resistance to *P. berghei* ANKA is not yet available. The cellular basis of the genetically determined development of cerebral pathology during this infection will be discussed in Section V.

B. GENETIC CONTROL OF RESISTANCE TO *PLASMODIUM YOELII*

Genetic control of resistance to *P. yoelii* 17XNL, which was originally characterized as being less virulent than *P. yoelii* 17XL, was established using the CXB series of RI mice which are derived from C57BL/6 and BALB/c progenitors.[54] For these studies, a cloned parasite preparation was used. Analysis of the level of resistance based on the time taken to reach peak parasitemia and the duration of the infection showed that C57BL/6 mice were susceptible and BALB/c mice were resistant. The pattern of infection in the RI strains derived from these two progenitors differed markedly from the two progenitor strains and did not correlate with the outcome of infection, thus, suggesting a multigeneic mode of inheritance of susceptibility to this parasite.

Differences among inbred strains of mice in the level of resistance to *P. yoelii* 17XNL have also been described by Taylor et al.[55,56] in terms of differences in (1) the course of parasitemia, (2) the antimalarial immunoglobulin isotypes produced, and (3) the malarial antigens against which antibodies are produced. The results of this series of experiments confirmed that BALB/c mice were more resistant than C57BL/6 mice. In addition to these strains, 16 other inbred and H-2 congenic strains and F1 hybrid combinations were typed for resistance by determining survival and the course of parasitemia. There was no correlation between the outcome of infection and the peak parasitemia or the length of the infection. However, inbred and H-2 congenic strains fell into three categories depending on the parasitemia on day 18. Group I (C3H, BALB/c, B10.BR) animals, which had completely cleared the parasite by day 18, and group II (NZB, DBA/2, B10.D2, B6.H.2k) animals, which had almost completely eliminated the parasite by day 18, exhibited high antibody titers. In particular, mice of these two groups which were relatively resistant to malaria produced high titers of IgG_2 and IgG_3 early in infection. Group III (AKR, C57BL/6, C57BL/10, B6TL$^+$) animals, which had ascending parasitemias on day 18 and occasional deaths, exhibited slower and diminished antibody production. The antibody responses of resistant (group I and II) and susceptible (group III) mice were found to differ to at least four antigens. Although the question of the mode of inheritance was not addressed in this study, the finding of a spectrum of responses among the three groups suggests multigenic control. H-2 linkage is suggested by the observation that several strain pair combinations which share the same H-2 haplotype, including H-2^d-compatible DBA and B10.D2 mice, are resistant.

Sayles and Wassom[57] examined inbred strains of mice for the level of resistance to infection to both *P. yoelii* 17XNL and *P. yoelii* 17XL. In agreement with the results of Taylor et al.[56] DBA/2 and H-2 congenic B10.D2 mice were found to be resistant to the 17XNL strain but susceptible to the 17XL strain. In contrast, C57BL/6 and C57BL/10 (genetically identical to B10.D2 except at H-2) were found to be susceptible to infection with *P. yoelii* 17XNL but resistant to *P. yoelii* 17XL. Susceptibility to the 17XNL strain was demonstrated to be expressed in C57BL-derived mice as an augmented production of antiparasite antibody as well as of immunoglobulins which bound to uninfected erythrocytes. These results suggest that susceptibility and not resistance, in contrast to the results of Taylor et al.,[56] was associated with a strong immune response to malaria antigens. Based on their observations, these investigators concluded that the ability to resist *P. yoelii* infections is dependent upon the genetic background of the host. Control is multigenic and requires H-2 linked genes as well as non-H-2 linked genes. Furthermore, the importance of the interplay between the host and the parasite genome is implicated by the finding that an individual inbred mouse strain may be resistant to infection with one malaria strain but susceptible to infection with a different strain of the same *Plasmodium* species.

C. GENETIC CONTROL OF RESISTANCE TO *PLASMODIUM CHABAUDI*

Differences among inbred strains of mice in response to infection with the AS strain of *P. chabaudi* were described by Eugui and Allison.[58,59] These investigators observed that C57BL/10 (H-2^b), CBA (H-2^k), and BALB/c (H-2^d) were resistant as determined by survival and development of a moderate, transient course of parasitemia while A/He (H-2^a) mice were susceptible with death occurring within 10 d with a fulminant parasitemia. C57BL-derived, H-2ᵃ congenic B10.A strain mice were found to be resistant suggesting that H-2 genes did not regulate resistance to *P. chabaudi* AS. A similar strain distribution of resistance was observed following infection with *Babesia microti*, a bovine hemoprotozoan parasite which can also infect mice. A/He mice developed a patent, persistent parasitemia and did not recover from this infection.

We confirmed and extended the observations of Eugui and Allison in a strain survey of 11 inbred strains of mice and 2 F1 hybrid combinations derived from a resistant and a susceptible parent.[60-64] Strains could be separated into two groups using survival as the criterion: (1) C57BL/6J, C57L/J, DBA/2J, CBA/J, and B10.A/SgSn mice were resistant while (2) A/J, DBA/1J, BALB/c, C3H/HeJ, AKR/J, and SJL/J mice were susceptible. Susceptible strains had a mean survival time of less than 10 d. F1 hybrids were found to be resistant indicating that resistance was dominant over susceptibility. Segregation analysis of backcross and F2 progeny derived from one of the most resistant strains, B10.A, and extremely susceptible A/J parental mice suggested that host resistance to *P. chabaudi* AS in this strain combination was genetically controlled by a single, dominant, non-H-2 linked gene called *Pchr*. Female mice were found to be more resistant than male mice, thus, suggesting that while inheritance of resistance was autosomal, expression of the trait was influenced by the sex of the host. The finding of unigenic control of resistance to infection with *P. chabaudi* AS was confirmed by analysis of over 20 AXB/BXA RI strains derived from susceptible A/J and resistant C57BL/6 progenitors. Resistance was found to be phenotypically expressed as moderate, transient parasitemia, marked splenomegaly, and an augmented erythropoietic response. The traits of resistance and the magnitude of splenomegaly were found to be genetically linked in the AXB/BXA RI strains. The *Pchr* gene did not influence the development of immunosuppression, a major complication of malaria in humans as well as experimental animals. Spleen cells from both C57BL/6 and A/J mice infected with *P. chabaudi* AS and immunized with sheep RBC had a greater than 50% depression in the number of direct plaque-forming cells. Based on the observations that superior levels of resistance in C57BL or DBA/2 mice were apparent within the 1st week after infection and that susceptible A/J mice could develop both nonspecific immunity following treatment with *Mycobacterium bovis*, BCG, or *Propionibacterium acnes* and specific immunity to reinfection after drug cure, we have hypothesized that genetically determined resistance to *P. chabaudi* AS is due to a mechanism of innate or natural resistance which is not yet defined but may involve splenic function.

Preliminary mapping performed by comparing the SDPs in the AXB/BXA RI strains for inheritance of *Pchr* with inheritance of known genetic loci showed linkage between *Pchr* and three different loci.[64] These loci were found to be peptidase-3 (*Pep-3*) which has been mapped to chromosome 1 of the mouse,[65] ornithine decarboxylase-2 (*Odc-2*) which has been mapped to chromosome 2,[66] and intestinal G protein (*Gi*) which has been mapped to chromosome 9.[67] The exact chromosomal location of *Pchr* can now be mapped using the tool of linkage analysis by typing animals of the susceptible background and F2 hybrid generations derived from A/J and C57BL/6 progenitors for the cosegregation of inheritance of *Pchr* and of the allelic forms of the marker genes, *Pep-3, Odc-2,* and *Gi*.

The results of studies by Borwell et al.,[68] who likewise examined the genetic control of resistance to *P. chabaudi* AS, demonstrated that C57BL/6 and DBA/2 strain mice had an intermediate level of resistance with DBA/2 being somewhat more susceptible. The percentage mortality was used to distinguish resistant and susceptible strains. C57BL/6 mice exhibited 74 to 85% mortality and DBA/2 mice exhibited 64 to 100% mortality. The outcome of infection appeared to be dependent on the source of the inoculum (*in vivo* passaged parasite material vs. frozen stabilate parasite material) and, in the case of some strains e.g., DBA/2, female mice exhibited superior resistance. Genetic analysis of 16 BXD RI strains derived from C57BL/6 and DBA/2 progenitors demonstrated that 4 RI strains were significantly more resistant than either of the progenitor strains. The conclusion reached by these investigators was that, in the strain pair combination of C57BL/6 and DBA/2 mice, resistance to *P. chabaudi* AS is genetically controlled by multiple genes. Preliminary mapping suggested that a major gene on chromosome 1, which was designated *Mal-1*, was responsible for the increased resistance of C57BL/6 mice. The existance of other genetic loci was postulated, including a locus mapping to chromosome 9. Borwell et al.[69] suggested that innate resistance, rather than acquired immunity, was responsible for the chromosome 1 gene controlled resistance to *P. chabaudi* AS.

Strain variation in resistance to infection with another *P. chabaudi* strain, *P. chabaudi adami*, was examined in C57BL and BALB/c mice and in mice of two CXB RI strains, CXBH and CXBJ by Hoffmann et al.[54] While differences were apparent among these strains in response to infection with *P. yoelii* 17XNL, the strains all exhibited a similar course and peak of parasitemia following infection with *P. chabaudi adami*. No deaths were observed in any of the strains. These results indicate that resistance to *P. chabaudi adami* does not cosegregate with resistance to *P. yoelii* 17XNL. These results confirm experiments performed by these same investigators using immunodeficient mice which demonstrated that different host resistance mechanisms are required for different parasites.[39] In addition, these results support experiments performed by Sayles and Wassom[57] in which the same inbred mouse strain was infected with different strains of *P. yoelii* 17X by demonstrating that host resistance mechanisms are dependent on not only the genetic background of the host but also on the *Plasmodium* species and strain.

V. GENETIC CONTROL OF MALARIA-ASSOCIATED PATHOLOGY

Much of the pathology associated with malaria appears to be the result of host immune responses, particularly cell-mediated responses, against parasite antigens (see Chapter 6). The major manifestations of what appear to be aberrant immune responses to malaria are cerebral malaria, thrombocytopenia, glomerulonephritis, and hyperreactive malarial splenomegaly (HMS), also called tropical splenomegaly syndrome (TSS). Genetic factors have been implicated to play a role in the pathogenesis of cerebral malaria and HMS. As discussed below, evidence supporting the role of the genetic background of the host in the development of cerebral malaria was derived from an experimental, murine model while that for HMS came from epidemiological studies.

A. CEREBRAL MALARIA
Cerebral malaria is a major and often fatal complication of human malaria due to *P. falciparum*, particularly in young children in hyperendemic areas.[69] The pathogenesis of cerebral malaria is only beginning to be understood as a result of studies performed using the murine model of infection with *P. berghei* ANKA.[70-74] Infection with this strain of *P. berghei* does not completely reproduce the human disease but does produce several features of the human condition, including the neurological symptoms.[70] As discussed in Section II.A, differences in response to infection with *P. berghei* ANKA are apparent among inbred mice. A survey of six inbred strains for resistance to cerebral complications demonstrated that DBA/2, BALB/c, and C3H mice are resistant while C57BL/6 and CBA mice are extremely susceptible. For example, 80% of CBA mice exhibited neurological symptoms 6 to 14 d after infection with a cumulative mortality of 90% within 15 d. One strain, A/J, was found have an intermediate level of resistance. The role of H-2 genes was not addressed in these studies.

The cellular basis of cerebral malaria was found to be dependent on L3T4+ T cells.

Studies by Grau et al.[72-74] using susceptible CBA mice and resistant BALB/c mice demonstrated that the neurological symptoms and pathology associated with cerebral malaria were caused by the release of tumor necrosis factor (TNF) or cachectin from macrophages presumably activated by L3T4$^+$ T cells. Serum levels of TNF were found to be extremely high in CBA mice but not BALB/c mice following infection with *P. berghei* ANKA. Treatment with rabbit antibody to TNF protected infected CBA mice from cerebral malaria but did not modify the parasitemia. Establishment of the role of TNF in the pathology of cerebral malaria is an excellent example of the use of a model of genetically determined resistance to malaria as a tool for determining the underlying mechanism.

B. HYPERREACTIVE MALARIAL SPLENOMEGALY

The major symptoms associated with HMS comprise a spectrum of symptoms which are not part of the normal immune response to malaria and which may include persistent splenomegaly, hepatic sinusoidal lymphocytosis with hyperplasia of the Kupffer cells, polyclonal macroglobulinaemia, high levels of immunofluorescent antimalarial antibody, cryoglobulinaemia, low C3 levels, red cell agglutinins in the sera or on erythrocytes, and the presence of rheumatoid factor.[75,76] HMS is thought to result from a prolonged stimulation of the reticuloendothelial elements of the spleen and liver by circulating immune complexes. Thus, it appears to be an immune complex disease but the mechanism is still unclear. Of interest here are the numerous observations of a genetic predisposition to the syndrome. For example, in Papua New Guinea, HMS is common in some areas but not in others with the same exposure to malaria.[77] In tropical Africa, HMS is generally uncommon but its presence has been described among immigrants to the area and familial aggregation of cases has been noted.[78] In several countries, the incidence of HMS was found to be very frequent among Fulanis.[79,80] An association between the HLA-DR2 and severe HMS has been reported in Papua New Guinea.[76] Unfortunately, there is not an experimental animal model to determine either the genetic basis or the underlying mechanism leading to HMS.

VI. GENETIC CONTROL OF THE IMMUNE RESPONSE TO SYNTHETIC *P. FALCIPARUM* SPOROZOITE VACCINES

The demonstration that monoclonal antibodies directed against the surface protein, the circumsporozoite (CS) protein, of malarial sporozoites could block infection suggested the CS protein as a potential malaria vaccine.[81] Characterization of the CS protein of *P. falciparum* showed that the immunodominant epitope consists of a tetrapeptide (Asn-Ala-Asn-Pro) repeated 37 times.[82,83] The majority of antibodies capable of reacting with sporozoites recognized this repeating epitope.[84] Both synthetic and recombinant peptides of the repetitive sequence of the *P. falciparum* CS protein have been produced and their immunogenicity has been examined in human volunteers and experimental animals.[33,85-89]

The influence of the genetic background of the host on responsiveness to the T cell epitope of the recombinant and synthetic peptides has been observed independently in two laboratories. Among H-2 congenic mouse strains, high titers of antibodies against the repetitive peptides in the absence of a carrier protein were produced only by mice which were of the H-2^b haplotype.[33,89] Analysis of H-2 recombinant and mutant mice by Good et al.[33] showed that the response was I-A linked (Table 3). As demonstrated by Del Giudice et al.,[89] the antibody response in nonresponder mice could be induced by immunization with the peptide coupled to KLH as a carrier protein. It is possible that human Ir genes likewise contributed to the poor immunogenicity of recombinant and synthetic sporozoite vaccine candidates in volunteers during human trials.[87,88]

An analogous situation may occur in humans during a natural infection. The use of an enzyme-linked immunosorbent assay, which employed the synthetic peptide, to determine

TABLE 3
Ir Gene Control of Response in Mice to the *P. falciparum* Sporozoite Recombinant Vaccine

Strain	H-2 alleles								Production of Anti-NANP IgG[a]
	K	A	B	J	E	C	S	D	
B10.S	s	s	s	s	s	s	s	s	−
B10.BR	k	k	k	k	k	k	k	k	−
B10.HTT	s	s	s	s	k	k	k	d	−
B10.A (5R)	b	b	b	k	k	d	d	d	+
B10	b	b	b	b	b	b	b	b	+
B10.D2	d	d	d	d	d	d	d	d	−
B10.MBR	b	k	k	k	k	k	k	q	−
B10.A (4R)	k	k	b	b	b	b	b	b	−

[a] NANP = tetrapeptide (Asn-Ala-Asn-Pro) of *P. falciparum* CS protein.

Adapted from Good, M. F., Berzofsky, J. A., Malo, W. L., Hayashi, Y., Fujii, N., Hockmeyer, W. T., and Miller, L. H., *J. Exp. Med.*, 164, 655, 1986. With permission.

the antibody level against the repetitive tetrapeptide of the CS protein of *P. falciparum* revealed that, at 10 years of age, about half the children living in a malaria endemic area in Tanzania did not develop such antibodies.[90] There were also significant differences in antibody levels against the repeat epitope among children living in different households but with similar exposure to malaria. The authors of this study concluded that genetic factors (possibly Ir genes) may influence the ability of an individual to respond to the repetitive epitope of the sporozoite CS protein during a natural infection. Similarly, in Gambia, human Ir genes may have contributed to the poor immunogenicity of the CS protein during natural infection.[91]

These observations, made in both humans and experimental animals, thus, emphasize the role that the genetic background of the host may play in the development of an immune response against the repetitive epitope whether it is presented as part of the sporozoite or as recombinant or synthetic peptides. It would appear to be imperative that the differential responsiveness based on the genetic background of the individual is considered in the design of an effective and safe malaria vaccine. This topic has recently been considered in depth in reviews by Grau et al.[74] and by Good et al.[92]

VII. CONCLUSION

Thus, the genetic background of the host appears to influence the response to malaria at several facets of host parasite interactions. This influence extends from the ability of the host to mount an immune response to sporozoites following the bite of an infected *Anopheles* mosquito through the invasion and multiplication of merozoites within host erythrocytes to the ensuing immunopathology. The outcome of these events appears to be modulated by host responses which are dependent on innate resistance as well as acquired immune mechanisms. Further study of genetic control of host resistance to malaria may result in the fundamental knowledge necessary for the manipulation of the host parasite relationship in favor of the host. This area may represent a previously unrecognized strategy for preventing and controlling malaria in human populations.

ACKNOWLEDGMENTS

This work was supported by grant number MA-7785 from the Medical Research Council of Canada. I would like to thank Dr. Michael F. Good for his helpful discussion and criticism during the preparation of this manuscript and Mrs. Mary Bergin for her excellent assistance in preparing the manuscript.

REFERENCES

1. **Miller, L. H., Mason, S. J., Clyde, D. F., and McGinniss, M. H.,** The resistance factor to *Plasmodium vivax* in blacks. The Duffy-blood-group genotype, FyFy, *N. Engl. J. Med.,* 295, 302, 1976.
2. **Boyd, M. F. and Stratman-Thomas, W. K.,** Studies on benign tertian malaria. IV. On the refractoriness of Negroes to inoculation with *Plasmodium vivax, Am. J. Hyg.,* 18, 485, 1933.
3. **Bray, R. S.,** The susceptibility of Liberians to the Madagascar strain of *Plasmodium vivax, J. Parasitol.,* 44, 371, 1958.
4. **Miller, L. H., Mason, S. J., Dvorak, J. A., McGinniss, M. H., and Rothman, I. K.,** Erythrocyte receptors for *Plasmodium knowlesi* malaria: Duffy blood group determinants, *Science,* 189, 561, 1975.
5. **Miller, L. H., Dvorak, J. A., Shiroishi, T., and Durocher, J. R.,** Influence of erythrocyte membrane components on malaria merozoite invasion, *J. Exp. Med.,* 138, 1597, 1973.
6. **Miller, L. H., Haynes, J. D., McAuliffe, F. M., Shiroishi, T., Durocher, J. D., and McGinniss, M. H.,** Evidence for differences in erythrocyte surface receptors for the malarial parasites, *Plasmodium falciparum* and *Plasmodium knowlesi, J. Exp. Med.,* 146, 277, 1977.
7. **Perkins, M.,** Inhibitory effects of erythrocyte membrane proteins on the *in vitro* invasion of the human malarial parasite (*Plasmodium falciparum*) into its host cell, *J. Cell. Biol.,* 90, 563, 1981.
8. **Pasvol, G., Wainscoat, J. S., and Weatherall, D. J.,** Erythrocytes deficient in glycophorin resist invasion by the malarial parasite *Plasmodium falciparum, Nature,* 297, 64, 1982.
9. **Serjeantson, S., Bryson, K., Amato, D., and Baboona, D.,** Malaria and hereditary ovalocytosis, *Hum. Genet.,* 37, 161, 1977.
10. **Booth, P. B., Serjeantson, S., Woodfield, D. G., and Amato, D.,** Selective depressions of blood group antigens associated with hereditary ovalocytosis among Melanesians, *Vox Sang.,* 32, 99, 1977.
11. **Kidson, C., Lamont, G., Saul, A., and Nurse, G. T.,** Ovalocytic erythrocytes from Melanesians are resistant to invasion by malaria parasites in culture, *Proc. Natl. Acad. Sci. U.S.A.,* 78, 5829, 1981.
12. **Hadley, T., Saul, A., Lamont, G., Hudson, D. E., Miller, L. H., and Kidson, C.,** Resistance of Melanesian elliptocytes (ovalocytes) to invasion by *Plasmodium knowlesi* and *Plasmodium falciparum* malaria parasites *in vitro, J. Clin. Invest.,* 71, 780, 1983.
13. **Breuer, W. V.,** How the malaria parasite invades its host cell, the erythrocyte, *Int. Rev. Cytol.* 96, 191, 1985.
14. **Allison, A. C.,** Protection afforded by sickle-cell trait against subtertian malarial infection, *Br. Med. J.,* 1, 290, 1954.
15. **Edington, G. M. and Watson-Williams, E. J.,** Sickling, haemoglobin C, glucose-6-phosphate dehydrogenase deficiency and malaria in Western Nigeria, in *Abnormal Haemoglobins in Africa,* Jonxis, J. H. P., Ed., Blackwell Scientific, Oxford, 1965, 393.
16. **Gilles, H. M., Fletcher, K. A., Hendrickse, R. G., Linder, R., Raddy, S., and Allan, N.,** Glucose-6-phosphate dehyrogenase deficiency, sickling, and malaria in African children in South Western Nigeria, *Lancet,* 1, 138, 1967.
17. **Friedman, M. J.,** Erythrocytic mechanism of sickle cell resistance to malaria, *Proc. Natl. Acad. Sci. U.S.A.,* 75, 1994, 1978.
18. **Pasvol, G., Weatherall, D. J., and Wilson, R. J. M.,** Cellular mechanism for the protective effect of haemoglobin S against *P. falciparum* malaria, *Nature,* 274, 701, 1978.
19. **Friedman, M. J. and Trager, W.,** The biochemistry of resistance to malaria, *Sci. Am.,* 244, 154, 1981.
20. **Livingstone, F. B.,** Malaria and human polymorphisms, *Annu. Rev. Genet.,* 5, 33, 1971.
21. **Lachant, N. A. and Tanaka, K. R.,** Impaired antioxidant defense in hemoglobin E-containing erythrocytes: a mechanism protective against malaria?, *Am. J. Hematol.,* 26, 211, 1987.
22. **Siniscalo, M., Bernini, L., Filippi, G., Latte, B., Meera Khan, P., Piomelli, S., and Rattazzi, M.,** Population genetics of hemoglobin variants, thalassemia and glucose-6-phosphate dehydrogenase deficiency, with particular reference to the malaria hypothesis, *Bull. W.H.O.,* 34, 378, 1966.
23. **Flint, J., Hill, V. S., Bowden, D. K., Oppenheimer, S. J., Sill, P. R., Serjeantson, S. W., Bana-Koiri, J., Bhatia, K., Alpers, M. P., Boyce, A. J., Weatherall, D. J., and Clegg, J. B.,** High frequencies of alpha-thalassemia are the result of natural selection by malaria, *Nature,* 321, 744, 1986.
24. **Pasvol, G., Weatherall, D. J., Wilson, R. J. M., Smith, D. H., and Gilles, H. M.,** Fetal hemoglobin and malaria, *Lancet,* 1, 1269, 1976.

25. **Allison, A. C.**, Glucose-6-phosphate dehydrogenase deficiency in red blood cells of East Africans, *Nature*, 185, 531, 1960.
26. **Allison, A. C. and Clyde, D. F.**, Malaria in African children with deficient erythrocyte glucose-6-phosphate dehydrogenase, *Br. Med. J.*, 1, 1345, 1961.
27. **Roth, E. F., Raventos-Suarez, C., Finaldi, A., and Nagel, R. L.**, Glucose-6-phosphate dehydrogenase deficiency inhibits *in vivo* growth of *Plasmodium falciparum*, *Proc. Natl. Acad. Sci. U.S.A.*, 80, 298, 1983.
28. **Friedman, M. J.**, Oxidant damage mediates variant red cell resistance to malaria, *Nature*, 280, 245, 1979.
29. **Etkin, N. L. and Eaton, J. W.**, Malaria-induced erythrocyte oxidant sensitivity, in *Erythrocyte Structure and Function*, Brewer, G. J., Ed., Alan R. Liss, New York, 1975, 219.
30. **Janney, S. K., Joist, J. H., and Fitch, C. D.**, Excess release of ferriheme in G6PD-deficient erythrocytes: possible cause of hemolysis and resistance to malaria, *Blood*, 67, 331, 1987.
31. **Osoba, D., Dick, H. M., Voller, A., Goosen, T. J., Goodsen, T., Draper, C. C., and de The, G.**, Role of the HLA complex in the antibody response to malaria under normal conditions, *Immunogenetics*, 8, 323, 1979.
32. **Good, M. F., Maloy, W. L., Lunde, M. N., Margalit, H., Cornette, J. L., Smith, G. L., Moss, B., Miller, L. H., and Berzofsky, J. A.**, Construction of synthetic immunogen: the use of new T-helper epitope on malaria circumsporozoite protein, *Science*, 235, 1059, 1987.
33. **Good, M. F., Berzofsky, J. A., Malo, W. L., Hayashi, Y., Fujii, N., Hockmeyer, W. T., and Miller, L. H.**, Genetic control of the immune response in mice to a *Plasmodium falciparum* sporozoite vaccine. Widespread non-responsiveness to single malaria T epitope in highly repetitive vaccine, *J. Exp. Med.*, 164, 655, 1986.
34. **Ott, K.**, Influence of reticulocytosis on the course of infection of *Plasmodium chabaudi* and *P. berghei*, *J. Protozool.*, 15, 365, 1968.
35. **Viens, P., Chevalier, J. L., Sonea, S., and Yoeli, M.**, The effect of reticulocytosis on *Plasmodium vinckei* infection in white mice. Action of phenylhydrazine and of repeated bleedings, *Can. J. Microbiol.*, 17, 257, 1971.
36. **Weinbaum, F. I., Evans, C. B., and Tigelaar, R. E.**, Immunity to *Plasmodium berghei yoelii* in mice. I. The course of infection in T cell and B cell deficient mice, *J. Immunol.*, 117, 1999, 1976.
37. **Jayawardena, A. N., Targett, G. A. T., Carter, R. L., Leuchars, E., and Davies, A. J. S.**, The immunological response of CBA mice to *P. yoelii*. I. General characteristics, the effects of T-cell deprivation and reconstitution with thymus grafts, *Immunology*, 32, 849, 1977.
38. **Taylor, D. W., Bever, C. T., Rollwagen, F. M., Evans, C. B., and Asofsky, R.**, The rodent malaria parasite *Plasmodium yoelii* lacks both types 1 and 2 T-independent antigens, *J. Immunol.*, 128, 1854, 1982.
39. **Grun, J. L. and Weidanz, W. P.**, Immunity to *Plasmodium chabaudi adami* in the B-cell-deficient mouse, *Nature*, 290, 143, 1981.
40. **Cox, F. E. G.**, Protective immunity between malaria parasites and piroplasms in mice, *Bull. W.H.O.*, 43, 325, 1970.
41. **Greenberg, J., Nadel, E. M., and Coatney, G. R.**, The influence of strain, sex and age of mice in infection with *Plasmodium berghei*, *J. Infect. Dis.*, 93, 96, 1953.
42. **Greenberg, J., Nadel, E. M., and Coatney, G. R.**, Differences in survival of several inbred strains of mice and their hybrids infected with *Plasmodium berghei*, *J. Infect. Dis.*, 95, 114, 1954.
43. **Nadel, E. M., Greenberg, J., and Coatney, G. R.**, Increased resistance to malaria in certain inbred mice, their hybrids and backcrosses, *Am. J. Pathol.*, 30, 658, 1954.
44. **Nadel, E. M., Greenberg, J., Jay, G. E., and Coatney, G. R.**, Backcross studies on the genetics of resistance to malaria in mice, *Genetics*, 40, 620, 1955.
45. **Greenberg, J.**, Differences in the course of *Plasmodium berghei* infections in some hybrid and backcross mice, *Am. J. Trop. Med. Hyg.*, 5, 19, 1956.
46. **Greenberg, J. and Kendrick, L. P.**, Parasitemia and survival in inbred strains of mice infected with *Plasmodium berghei*, *J. Parasitol.*, 43, 413, 1957.
47. **Greenberg, J. and Kendrick, L. P.**, Some characteristics of *Plasmodium berghei* passed within inbred strains of mice, *J. Parasitol.*, 43, 420, 1957.
48. **Greenberg, J. and Kendrick, J. L.**, Parasitemia and survival in mice infected with *Plasmodium berghei*. Hybrids between Swiss (high parasitemia) and STR (low parasitemia) mice, *J. Parasitol.*, 44, 492, 1958.
49. **Greenberg, J. and Kendrick, L. P.**, Resistance to malaria in hybrids between Swiss and certain other strains of mice, *J. Parasitol.*, 45, 263, 1959.
50. **Eling, W., van Zon, A., and Jerusalem, C.**, The course of a *Plasmodium berghei* infection in six different mouse strains, *Z. Parasitenkd.*, 54, 29, 1977.
51. **van Zon, A., Eling, W., and Jerusalem, C.**, Histo-and immunopathology of a malaria (*Plasmodium berghei*) infection in mice, *Isr. J. Med. Sci.*, 14, 6, 1978.

52. **Contreras, C. E., June, C. H., Perrin, L. H., and Lambert, P. H.,** Immunopathological aspects of *Plasmodium berghei* infection in five strains of mice. I. Immune complexes and other serological features during the infection. *Clin. Exp. Immunol.,* 42, 403, 1980.

53. **Mackey, L. J., Hockmann, A., June, C. H., Contreras, C. E., and Lambert, P. H.,** Immunopathological aspects of *Plasmodium berghei* infection in five strains of mice. II. Immunopathology of cerebral and other tissue lesions during the infection, *Clin. Exp. Immunol.,* 42, 412, 1980.

54. **Hoffmann, E. J., Weidanz, W. P., and Long, C. A.,** Susceptibility of CXB recombinant inbred mice to murine plasmodia, *Infect. Immun.,* 43, 981, 1984.

55. **Taylor, D. W., Evans, C. B., and Asofsky, R.,** Antigenic responsiveness and kinetics of antibody production differ in inbred strains of mice infected with the malarial parasite, *Plasmodium yoelii,* in *Genetic Control of Host Resistance to Infection and Malignancy,* Skamene, E., Ed., Alan R. Liss, New York, 1985, 545.

56. **Taylor, D. W., Pacheco, E., Evans, C. B., and Asofsky, R.,** Inbred mice infected with *Plasmodium yoelii* differ in their antimalarial immunoglobulin isotype response, *Parasite Immunol.,* 10, 33, 1988.

57. **Sayles, P. C. and Wassom, D. L.,** Immunoregulation in murine malaria: susceptibility of inbred mice to infection with *Plasmodium yoelii* depends on the dynamic interplay of host and parasite genes, *J. Immunol.,* 141, 241, 1988.

58. **Eugui, E. M. and Allison, A. C.,** Malaria infection in different strains of mice and their correlation with natural killer activity, *Bull. W.H.O.,* 57 (Suppl. 1), 231, 1979.

59. **Eugui, E. M. and Allison, A. C.,** Differences in susceptibility of various mouse strains to haemoprotozoan infections: possible correlation with natural killer activity, *Parasite Immunol.,* 2, 277, 1980.

60. **Stevenson, M. M., Lyanga, J. J., and Skamene, E.,** Murine malaria: genetic control of resistance to *Plasmodium chabaudi, Infect. Immun.,* 38, 80, 1982.

61. **Stevenson, M. M. and Skamene, E.,** Murine malaria: resistance of AXB/BXA recombinant inbred mice to *Plasmodium chabaudi, Infect. Immun.,* 47, 452, 1985.

62. **Stevenson, M. M.,** Genetic control of resistance in mice to *Plasmodium chabaudi,* in *Genetic Control of Host Resistance to Infection and Malignancy,* Skamene, E., Ed., Alan R. Liss, New York, 1985, 531.

63. **Stevenson, M. M. and Skamene, E.,** Modulation of primary antibody responses to sheep erythrocytes in *Plasmodium chabaudi*-infected resistant and susceptible mouse strains, *Infect. Immun.,* 54, 600, 1986.

64. **Stevenson, M. M., Nesbitt, M., and Skamene, E.,** Chromosomal location of the gene determining resistance to *Plasmodium chabaudi,* in *Current Topics in Microbiology and Immunology,* Mock, B. and Potter, M., Eds., Springer-Verlag, New York, 1988, 325.

65. **Lewis, W. H. and Truslove, G. M.,** Electrophoretic heterogeneity of mouse erythrocyte peptidase, *Biochem. Genet.,* 3, 493, 1969.

66. **Nesbitt, M. N.,** unpublished observation.

67. **Cohen, V.,** unpublished observation.

68. **Borwell, P., Holmquist, B. G. F., Cattan, A., Lundin, L.-G., Hakansson, E. M., and Wigzell, H.,** Genetics of resistance to malaria in the mouse. I. Association of innate resistance to *Plasmodium chabaudi* with chromosome 1 markers, in *Experimental Bacterial and Parasitic Infections,* Keusch, G. and Wadstrom T., Eds., Elsevier, New York, 1983, 355.

69. **Jerusalem, C., Polder, T., Wijers-Rouw, M., Heinen, U., Eling, W., Osunkoya, B. O., and Trinh, P.,** Comparative clinical and experimental study on the pathogenesis of cerebral malaria, *Contrib. Microbiol. Immunol.,* 7, 130, 1983.

70. **Finley, R. W., Mackey, L. J., and Lambert, P. H.,** Virulent *P. berghei* malaria: prolonged survival and decreased cerebral pathology in cell-deficient nude mice, *J. Immunol.,* 129, 2213, 1982.

71. **Grau, G. E. and Lambert, P. H.,** personal communication.

72. **Grau, G. E., Piguet, P. F., Engers, H. D., Louis, J. A., Vassalli, P., and Lambert, P. H.,** L3T4[+] T lymphocytes play a major role in the pathogenesis of murine cerebral malaria, *J. Immunol.,* 137, 2348, 1986.

73. **Grau, G. E., Fajardo, L. F., Piguet, P. F., Allet, B., Lambert, P. H., and Vassalli, P.,** Tumor necrosis factor (cachectin) as an essential mediator in murine cerebral malaria, *Science,* 237, 1210, 1987.

74. **Grau, G. E., Del Giudice, G., and Lambert, P. H.,** Host immune response and pathological expression in malaria: possible implications for malaria vaccines, *Parasitology,* 94, S123, 1987.

75. **Crane, G. G.,** The serology of tropical splenomegaly syndrome and its relationship to malaria, in *The Role of the Spleen in the Immunology of Parasitic Diseases,* Tropical Disease Research Series, Schwabe and Co., Basel, 1979, 245.

76. **Crane, G. G.,** Hyperreactive malarious splenomegaly (tropical splenomegaly syndrome), *Parasitol. Today,* 2, 4, 1986.

77. **Crane, G. G. and Pryor, D. S.,** Malaria and tropical splenomegaly syndrome in New Guinea, *Trans. R. Soc. Trop. Med. Hyg.,* 65, 315, 1971.

78. **Ziegler, J. L. and Stuiver, P. C.,** Tropical splenomegaly syndrome in a Rwandan kindred in Uganda, *Br. Med. J.,* iii, 79, 1972.

79. Bryceson, A. D. M., Fakunle, Y. M., Fleming, A. F., Crane, G., Hutt, M. S. R., DeCock, K. M., Greenwood, B. M., Marsden, P., and Rees, P., Malaria and splenomegaly, *Trans. R. Soc. Trop. Med. Hyg.*, 77, 879, 1983.

80. Greenwood, B. M., Groenendaal, F., Bradley, A. K., Greenwood, A. M., Shenton, F., and Tulloch, S., Ethnic differences in the prevalence of splenomegaly and malaria in The Gambia, *Ann. Trop. Med. Parasitol.*, 81, 345, 1987.

81. Potocnjak, P., Yoshida, P., Nussenzweig, R. S., and Nussenzweig, V. J., Monovalent fragments (Fab) of monoclonal antibodies to a sporozoite surface antigen (Pb44) protect mice against malarial infection, *J. Exp. Med.*, 151, 1504, 1980.

82. Dame, J. B., Williams, J. L., McCutchan, T. F., Weber, J. L., Wirtz, R. A., Hockmeyer, W. T., Maloy, W. L., Haynes, J. D., Schneider, I., Roberts, D., Sanders, G. S., Reddy, E. P., Diggs, C. L., and Miller, L. H., Structure of the gene encoding the immunodominant surface antigen on the sporozoite of the human malaria parasite *Plasmodium falciparum*, *Science*, 225, 593, 1984.

83. Enea, V., Ellis, J., Zavala, F., Arnot, D. E., Asanavich, A., Masuda, A., Quakyi, I., and Nussenzweig, R. S., DNA cloning *Plasmodium falciparum* circumsporozoite gene: amino acid sequence of repetitive epitope, *Science*, 225, 628, 1984.

84. Zavala, F., Cochrane, A. H., Nardin, E. H., Nussenzweig, R. S., and Nussenzweig, V., Circumsporozoite proteins of malaria parasites contain a single immunodominant region with two or more identical epitopes, *J. Exp. Med.*, 157, 1947, 1983.

85. Etlinger, H. M., Felix, A. M., Gillessen, D., Heimer, E. P., Just, M., Pink, J. R. L., Sinigaglia, F., Sturchler, D., Takas, B., Trziciak, A., and Matile, H., Assessment in humans of a synthetic peptide-based vaccine against the sporozoite stage of the human malaria parasite, *Plasmodium falciparum*, *J. Immunol.*, 140, 626, 1988.

86. Young, J. F., Hockmeyer, W. T., Gross, M., Ballou, W. R., Wirtz, R. H., Trosper, J. H., Beaudoin, R. L., Hollingdale, M. R. Miller, L. H., Diggs, C. L., and Rosenberg, M., Expression of *P. falciparum* circumsporozoite derivatives in *E. coli* for development of human malaria vaccine, *Science*, 228, 958, 1985.

87. Ballou, W. R., Rothbard, J., Wirtz, R. A., Gordon, D. M., Williams, J. S., Gore, R. W., Schneider, I., Hollingdale, M. R., Beaudoin, R. L., Maloy, W. L., Miller, L. H., and Hockmeyer, W. T., Immunogenicity of synthetic peptides from circumsporozoite protein of *Plasmodium falciparum*, *Science*, 228, 996, 1985.

88. Herrington, D. A., Clyde, D. F., Losonsky, G., Cortesia, M., Davis, J., Murphy, J. R., Felix, A. M., Heimer, E. P., Gillessen, D., Nardin, E., Nussenzweig, R. S., Nussenzweig, V., Hollingdale, M. R., and Levine M. M., Safety and immunogenicity in man of a synthetic peptide tetanus toxoid conjugate malaria vaccine against *Plasmodium falciparum* sporozoites, *Nature*, 328, 257, 1987.

89. Del Giudice, G., Cooper, J. A., Merino, J., Verdini, A. S., Pessi, A., Togna, A. R., Engers, H. D., Corradin, G., and Lambert, P. H., The antibody response in mice to carrier-free synthetic polymers of *Plasmodium falciparum* circumsporozoite repetitive epitope is I-Aᵇ-restricted: possible implications for malaria vaccines, *J. Immunol.*, 137, 2952, 1986.

90. Del Giudice, G., Engers, H. D., Tougne, C., Biro, S. S. Weiss, N., Verdini, A. S., Pessi, A., Degremont, A. A., Freyvogel, T. A., Lambert, P. H., and Tanner, M., Antibodies to the repetitive epitope of *Plasmodium falciparum* circumsporoazoite protein in a rural Tanzanian community: a longitudinal study of 132 children, *Am. J. Trop. Med. Hyg.*, 36, 203, 1987.

91. Good, M. F., Pombo, D., Quakyi, I. S., Riley, E. M., Houghten, R. A., Menon, A., Alling, D. W., Berzofsky, J. A., and Miller, L. H., Human T cell recognition of the circumsporozoite protein of *Plasmodium falciparum*. Immunodominant T cell domains maps to the polymorphic regions of the molecule, *Proc. Natl. Acad. Sci. U.S.A.*, 85, 1199, 1988.

92. Good, M. F., Berzofsky, J. A., and Miller, L. H., The T-cell response to the malaria circumsporozoite protein: an immunological approach to vaccine development, *Annu. Rev. Immunol.*, 6, 663, 1988.

INDEX